Belarus Heat Tariff Reform
and Social Impact Mitigation

A WORLD BANK STUDY

Belarus Heat Tariff Reform
and Social Impact Mitigation

Fan Zhang and Denzel Hankinson

WORLD BANK GROUP

Contents

Map

Tables

Acknowledgments

This work was carried out under the direction of Qimiao Fan (Country Director for Belarus, Moldova, and Ukraine) and Ranjit Lamech (Practice Manager, Europe and Central Asia Region, Energy and Extractives Global Practice). The team thanks them for their generous guidance, insights, and encouragement.

This report was prepared by a team led by Fan Zhang. A large number of World Bank staff and consultants from the Energy and Extractives; Macroeconomics and Fiscal Management; Poverty; Social Protection and Labor; and Social, Urban, Rural, and Resilience Global Practices, as well as from the External Affairs Unit, provided substantial contributions. Vlad Grigoras and Julia Smolyar contributed to the social protection analysis. Sebastian Eckardt contributed to the fiscal analysis. Murat Alehodzhin and Irina Voitekhovitch prepared a detailed background study on energy efficiency. Ecaterina Canter, Izabela Leao, and Klavdiya Maksymenko conducted extensive work on the qualitative social impact analysis and stakeholder analysis. Corbett Grainger and Bonsuk Koo contributed to the quantitative social impact analysis. Irina Oleinik led the communication study and training. Yating Chuang, Bonsuk Koo, Deborah Ong, Karuna Phillips, and Andrew Schreiber provided valuable research assistance. Maryna Sidarenka and Irina Voitekhovitch worked tirelessly to obtain household survey and administrative data. Valdas Lukosevicius provided background studies on tariff setting methodologies and tariff reform experiences in the Baltic countries and Poland. Dianne Snyder edited the report. Rozena Serrano provided timely administrative assistance.

The study received substantial support from the Europe and Central Asia management team. The team thanks in particular Elisabeth Huybens in the Social Development Unit, Andrew Mason in the Social Protection Unit, and Carolina Sanchez-Paramo in the Poverty Unit. The team is also grateful to representatives of the joint working group from the Ministries of Finance, Economy, Energy, Housing and Utilities, Labor and Social Protection; as well as from Belenergo, Eltopgas, the Energy Efficiency Department, and the oblast executive committees, for their productive discussions and feedback throughout the project. The team is also grateful to the National Statistical Committee for providing essential data.

The report benefited greatly from Elena Klochan who worked tirelessly to coordinate with the government, provided important guidance on country issues, and supported the establishment of the joint working group. Pekka Salminen has nurtured and led the World Bank's engagement in the district heating sector in Belarus. The team is greatly indebted to Pekka for generously providing his guidance and insights on technical and policy issues of the sector. The peer reviewers for this report are Sameer Shukla, Jonathan Sinton, Emil Tesliuc, Maria Vagliasindi, and Yadviga Viktorivna. Ani Balabanyan and Caterina Laderchi carefully reviewed the report at multiple stages and provided comments on the substance and presentation of the report. The team thanks all of the reviewers for their valuable inputs and suggestions at various stages of the study. The team gives special thanks also to Nicolas Perrin and Michelle Rebosio, who supported and coordinated the qualitative field survey and the stakeholder analysis.

In addition to the contributors of the background papers and peer reviewers, many people provided helpful comments, suggestions, and other inputs along the way. The team thanks in particular Alejandro Cedno, Stephane Dahan, Uwe Deichmann, Sophia Georgieva, Sunil Khosla, Young Chul Kim, Kari Nyman, Ian Parry, Jas Singh, Claudia Vasquez Suarez, Tamara Sulukhia, Nithin Umapathi, and Heather Worley. The team apologizes to anyone inadvertently overlooked in these acknowledgments.

The team thanks ESMAP ABG and Subsidies Trust Fund and ECA PSIA Trust Fund for their generous financial support.

Abbreviations

CHP   combined heat and power plants
DH   district heating
EU   European Union
FGD   focus group discussion
GDP   gross domestic product
GoB   government of Belarus
H&U   Housing and Utility
IDI   in-depth interview
PWS   Program of Winter Surcharges

Executive Summary

Introduction

The government of Belarus (GoB) plans to increase district heating (DH) tariffs to cost-recovery levels and gradually phase out subsidies, replacing them with social assistance programs. Residential DH tariffs in Belarus are currently at roughly 10–21 percent of cost-recovery levels. DH subsidies are highly regressive, add costs to business, and create significant fiscal risks and macroeconomic vulnerabilities.

The purpose of this report is to analyze the social, sectoral, and fiscal impacts of the proposed tariff reform and to identify and recommend measures to mitigate adverse impacts of DH tariff increases on the households. The analysis shows the following:

- The burden of higher DH tariffs will fall most heavily on low-income groups.
- The current system of subsidies is unfair, benefitting wealthy customers more than the poor.
- Cross-subsidies undermine the competitiveness of industries in Belarus.
- Underpriced residential heat places an increasing fiscal burden on the GoB and risks macroeconomic instability.

The analysis shows that a negative social impact is manageable if a tariff increase is accompanied by countervailing measures to compensate for the loss of purchasing power, in particular of the poor, through targeted social assistance and energy efficiency programs. The reform is more likely to be successful if communication campaigns to address consumer concerns are carried out before significant price increases, and consumer engagement and monitoring systems are established. When tariff reform and mitigation measures are properly sequenced and coordinated, the reform will become more socially acceptable, consumers will benefit from better quality of services, the government will achieve positive fiscal savings, and the DH sector will become sustainable in the long term. A sustainable DH sector means the following:

- Financially viable DH service providers—Belenergo and ZhKH—that can afford to maintain and invest as much as is required to provide the services that the customers want

- The independence of DH service providers from excessive direct fiscal subsidies
- Well-targeted social assistance for customers struggling to afford the cost of heating.

Table I.1 summarizes the challenges facing the DH sector and the recommended policy options.

The rest of the report is organized as follows: Chapter 1 describes the GoB's plans for the sector. Chapter 2 analyzes the principal challenges in the sector that necessitate tariff reform. Chapter 3 discusses tariff reform options and the likely impact of pursuing each of these options. Chapter 4 concludes by recommending a reform action package that includes customer communication and engagement, social protection measures, and investments in energy efficiency. The appendices contain material supporting the analysis in each section.

Table I.1 Policy Matrix for Tariff and Subsidy Reform in District Heating

Challenges	Recommended measures			Expected impact
	2014–15	*2016–17*	*2018–20*	
Residential tariffs are well below cost of service and fiscal burden continues increasing	Achieve 30 percent cost-recovery for Belenergo and ZhKH by 2015	Gradually increase the cost-recovery for Belenergo and ZhKH		Belenergo and ZhkH can continue to provide reliable, good-quality service with limited fiscal impact on the government of Belarus
Possible customer resistance to tariff increases	Roll out consumer communication campaign Establish consumer monitoring mechanisms	Continue improving the transparence and accountability of utility services		Customer acceptance of tariff increases
Tariff increases will hurt the poor more than the rich	Better social protection measures established Provide preferential loan/grants to low-income households for demand-side energy efficiency Supply- and demand-side energy-efficiency measures implemented		Demand-side energy efficiency fully scaled up	Targeted relief for vulnerable customers; limited impact on affordability of heat supply

Source: State Program on the Energy Sector Development by 2016.

CHAPTER 1

What Are the Government's Plans for the Sector?

Sector Plans

The government of Belarus (GoB) has set national targets, planned investments, and continued to enact tariff reform in the district heating (DH) sector. The GoB's Strategy for Energy Potential Development sets national targets for the energy sector up until 2020. The overall objective of the strategy is to ensure Belarus's energy independence and promote the efficient use of energy resources. The GoB targets relevant to the DH sector include the following:

- Increasing the share of domestic fuel in the energy mix to 28–30 percent by 2015
- Increasing the share of domestic fuel in the energy mix to 32–34 percent by 2020 from 17 percent in 2010, thus reducing dependence on imported natural gas
- Reducing the share of natural gas in the energy balance to 64 percent in 2015 and to 55 percent by 2020
- Reducing the energy intensity of gross domestic product (GDP) by 50 percent by 2015 and 60 percent by 2020 (from 2005 levels)
- Phasing out subsidies and cross-subsidies
- Restructuring heat tariffs.

In line with this strategy, the GoB has enacted a number of DH sector-specific policies and laws subsequently described.

One of Belarus's richest natural resources is its forests, which cover 40 percent of the country. The GoB intends to increase the share of electricity and thermal energy generated from biomass to 14–15 percent so that, by 2020, at least 32 percent of the fuel used in boilers comes from locally sourced fuels.[1]

Table 1.1   National Cost-Recovery-Level Targets for the District Heating Sector, 2011–15

	2011 (actual)	2012 (actual)	2013	2014	2015
Cost-recovery rate for heat (distributed by Belenergo suppliers, %)	21.4	17.2	18.7	21	30
Prime cost of 1 Gcal of heat, BYR/Gcal	202,185.50	329,273.90	359,649.60	406,217.80	453,138.40

Source: State Program on the Energy Sector Development by 2016.

The GoB has also set national cost-recovery targets for the residential DH operations of Belenergo, a major heat provider in Belarus. Belenergo is expected to achieve 30 percent cost-recovery levels in its residential heating operations by 2015. In 2012, Belenergo's cost-recovery level for residential heat services was only 17.2 percent. Table 1.1 summarizes national cost-recovery targets for the DH sector.

At the municipal level, the GoB has enacted the Program for Housing and Utilities of the Republic of Belarus 2015, which aims to reduce heat losses by 6.7 percent in the heat network by 2016 by replacing old and inefficient heat network, introducing more energy-efficient generation facilities, reducing subsidies and cross-subsidies, and increasing the use of local fuels. From an organizational standpoint, the GoB also plans to centralize the DH sector, transferring municipal ownership to national ownership under Belenergo to extract efficiency gains.[2]

More recently, to simplify the system of cross-subsidies between the electricity and DH sectors and between residential and industrial customers, the GoB has phased out preferential heat tariffs for legal entities and individual entrepreneurs and is gradually increasing residential tariffs each financial quarter. These increases are indexed by the growth of household income, which does not exceed the growth of nominal wages. However, tariff increases for all energy utilities cannot increase by more than US$5 per year without approval from the president (Decree 550). As a mitigation measure, households in urban areas whose income on utilities exceeds 20 percent and those in rural areas whose income exceeds 15 percent will receive social assistance. Appendix A provides an overview of the DH sector in Belarus.

Notes

1. National Program of Local and Renewable Energy Sources Development for 2011–15.
2. Heat Supply Development Concept for the Period until 2020.

CHAPTER 2

Why Is Tariff Reform Necessary?

Tariff Reform

Residential tariffs for district heating (DH) are well below the cost of service in Belarus. Since 2003, production costs have risen sharply while the cost-recovery levels of residential heat service have dropped by 50 percent. Incremental increases in residential tariffs have been eroded by inflation and depreciation of the Belarusian ruble to the US dollar. Even if tariffs were increased by US$5 per year, the limit before presidential approval is necessary, and they would not meet the 30 percent cost-recovery target set by the government of Belarus (GoB). A system of subsidies and cross-subsidies between customer classes and between the electricity and the DH sectors have resulted in an increasing fiscal burden, which worsens as the cost of service continues to increase.

Residential tariffs are currently at roughly 10–21 percent of cost-recovery levels. The range depends on factors that include the size of the DH system, fuel used, efficiency of production, condition of the networks, and technical characteristics of the customer connection. Figure 2.1 shows how the cost-recovery levels of residential heat service have changed over time in Belarus.

Costs faced by suppliers of heating have risen substantially in recent years and are higher than the "economically efficient" level assessed by the Council of Ministers.[1] The cost of fuel for use in combined heat and power plants and boilers is the most significant cost faced by suppliers, not least because it is paid in US dollars. The price of importing natural gas from Russia has increased sharply in the past decade, from US$47/tcm in 2005 to US$263/tcm in 2011. Over the same period, the value of the ruble has fallen considerably. This is offset only slightly by the reduction in technical losses in the transmission and distribution systems—currently 16.3 percent for ZhKH and 10 percent for Belenergo. Figure 2.2 shows the rapid increase in the price of natural gas imports since 2005. The import price of natural gas accounts for roughly 60 percent of total heat production costs.

Figure 2.1   Declining Cost-Recovery Levels of Residential Heat Service, 2005–12

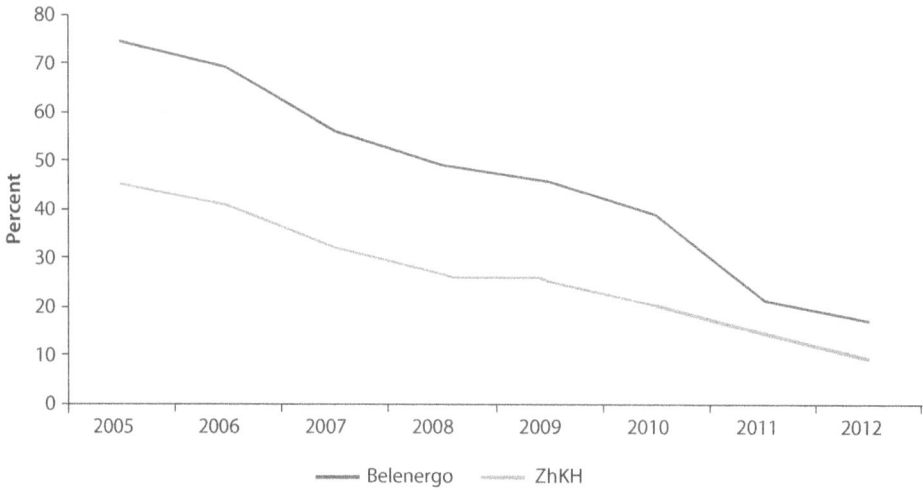

Source: Ministry of Energy, Ministry of Housing and Utilities.
Note: Belenergo and ZhKH are the two major district heating providers in Belarus. See appendix A for a background of the Belarusian district heating sector.

The cost of heat production and distribution by ZhKH is thought to be about double that of Belenergo. Contributing factors are (i) economies of scale (Belenergo serves customers in Minsk city and the oblast capitals, while ZhKH serves customers in smaller towns and rural areas), (ii) Belenergo's use of efficient combined heat and power plants (in contrast to the use of heat-only boilers by ZhKH), and (iii) the higher transmission and distribution losses in the ZhKH systems as a result of aging ZhKH assets. Figure 2.3 shows the increase in production costs for ZhKH and Belenergo, as compared to the increase in tariffs.

To cover costs in an environment where household tariffs are constrained, Belenergo and ZhKH have needed to make up the loss from supplying residential heat consumers from other sources. Belenergo, which on average achieves 17.2 percent cost recovery from residential heat consumers, does not receive state subsidies and so must make up the entire shortfall by cross-subsidization.[2] As a result, Belenergo's nonresidential energy consumers, mostly nonresidential electricity consumers, pay tariffs that are substantially above cost in order to keep heating prices low for residential consumers. Figure 2.4 compares industrial electricity tariffs with the cost of service of industrial electricity since 2005. It shows that Belenergo's industrial customers are paying a 50 percent premium on electricity to support underpriced residential heat.

The situation for ZhKH, whose cost recovery from residential heat sales is about 10 percent, is similar, except that ZhKH have compensated for the falling value of residential revenue with substantial increases in state subsidies, together with cross-subsidization from nonresidential consumers (figure 2.5).

Figure 2.2   Import Prices of Russian Natural Gas, 2005–12

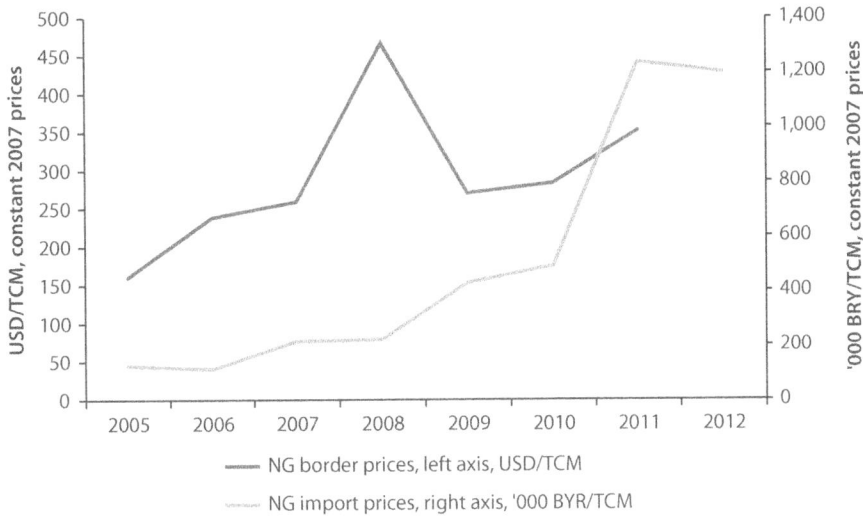

— NG border prices, left axis, USD/TCM

‑‑‑ NG import prices, right axis, '000 BYR/TCM

Source: Ministry of Energy, OECD.

Figure 2.3   Comparison of Tariffs and Production Costs of ZhKHs and Belenergo, 2005–12

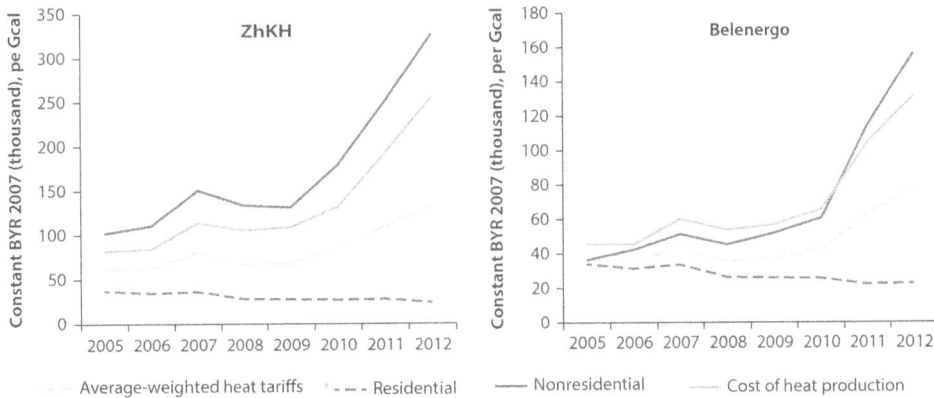

‑‑‑‑ Average-weighted heat tariffs   ‑ ‑ ‑ Residential   —— Nonresidential   ——— Cost of heat production

Source: Ministry of Energy, Ministry of Housing and Utilities.

By charging higher tariffs to nonresidential customers, cross-subsidies also impose an implicit tax on industries and could undercut the competitiveness of the economy. If industrial electricity tariffs are reduced to cost-recovery levels at about US 9.25 cents/kWh, the energy cost of manufacturing in Belarus could be reduced by roughly 24 percent, making it a lower-cost producer than the European Union (EU) average. Figure 2.6 compares the unit energy cost per US dollar of industrial value added with and without cross-subsidies with those of its neighboring countries.

Belarus Heat Tariff Reform and Social Impact Mitigation
http://dx.doi.org/10.1596/978-1-4648-0696-4

Figure 2.4 Industrial Electricity Tariffs, 2005–14

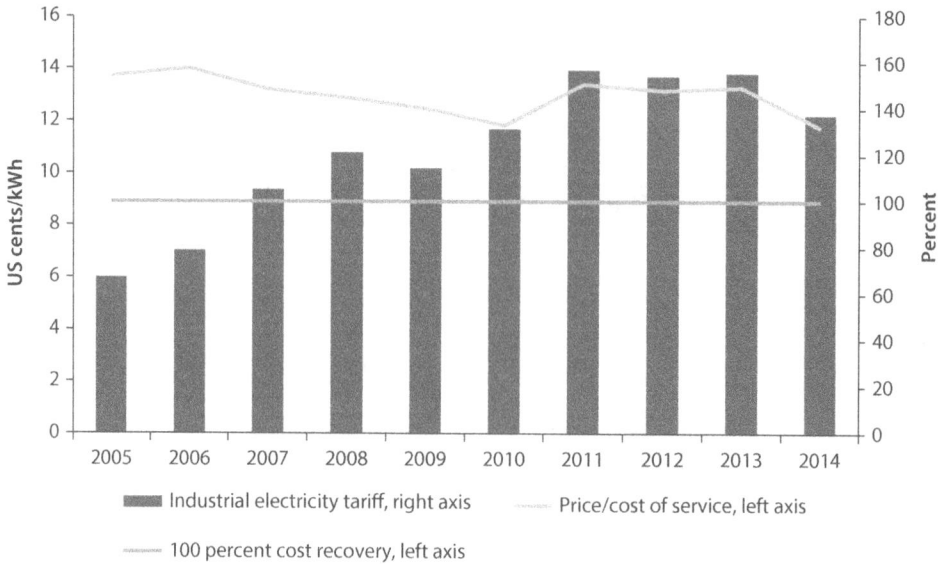

Legend:
Industrial electricity tariff, right axis — Price/cost of service, left axis
100 percent cost recovery, left axis

Source: Ministry of Energy, Ministry of Housing and Utilities.

Figure 2.5 Cross- and Direct Budgetary Subsidies to Residential District Heating, 2005–12

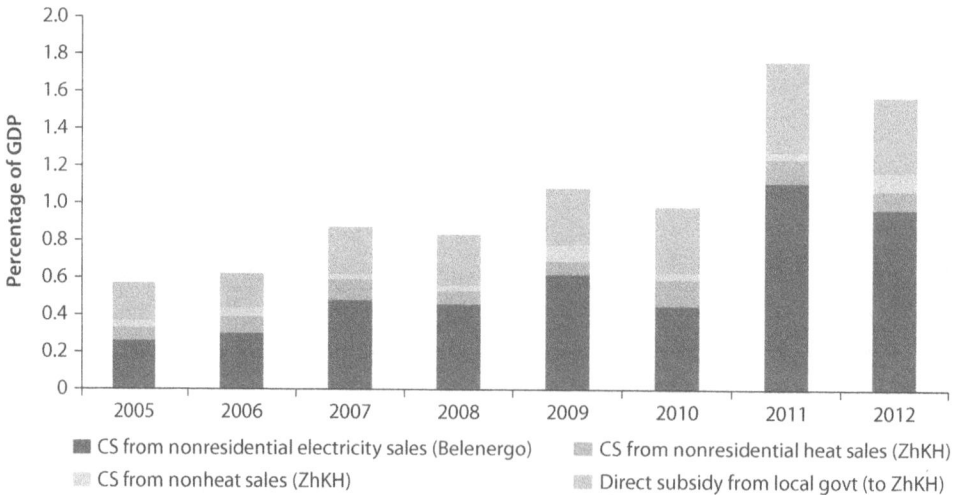

Legend:
CS from nonresidential electricity sales (Belenergo) — CS from nonresidential heat sales (ZhKH)
CS from nonheat sales (ZhKH) — Direct subsidy from local govt (to ZhKH)

Source: Ministry of Finance, Belenergo, ZhKH, and Beltopgas estimates.

Because electricity is required to produce and distribute goods from all sectors in an economy, an implicit tax on industrial electricity use will likely impact the price of nearly all goods and services. To the extent possible, firms facing an electricity tax will pass the tax burden on to consumers or other firms in the form of higher output prices.

Belarus Heat Tariff Reform and Social Impact Mitigation
http://dx.doi.org/10.1596/978-1-4648-0696-4

Figure 2.6   Unit Energy Cost of Manufacturing, by Country

$/manufacturing value added

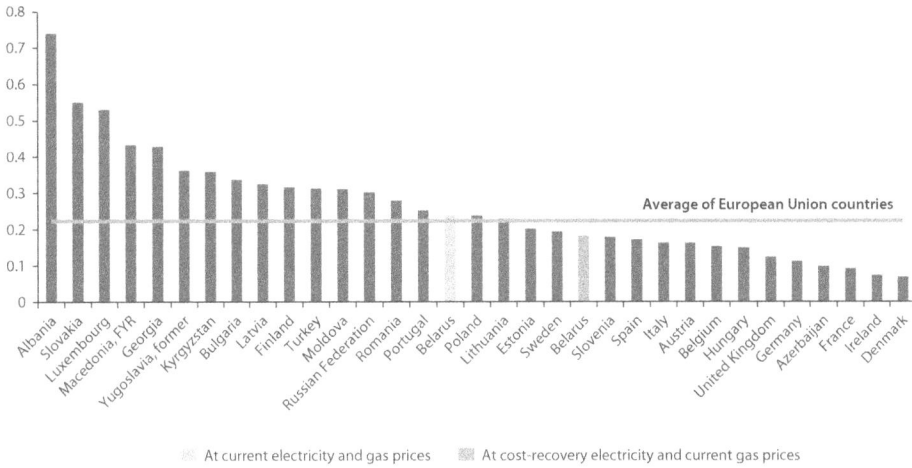

At current electricity and gas prices   At cost-recovery electricity and current gas prices

Source: Ministry of Economy.

Note: Energy cost of manufacturing in Belarus could be reduced by 24 percent if cross-subsidies are removed.

Figure 2.7   Output Price Increases from Imposing Implicit Electricity Tax on Industrial Consumers

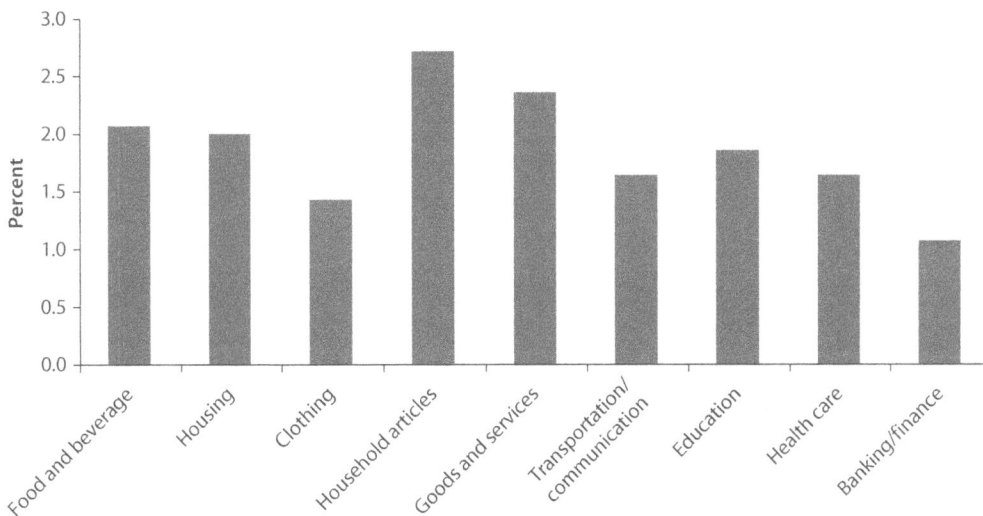

Source: Grainger, Zhang, and Schreiber 2015.

On the basis of an input–output analysis, figure 2.7 shows the increase in output prices across consumer expenditure categories as a result of the current high tax rates on industrial electricity use. The impacts range from a 1 percent increase in the output price in the banking and finance sector to an almost 3 percent increase in the output price for household articles. For food and

Figure 2.8   Expenditure Shares, by Consumption Category and Income Decile

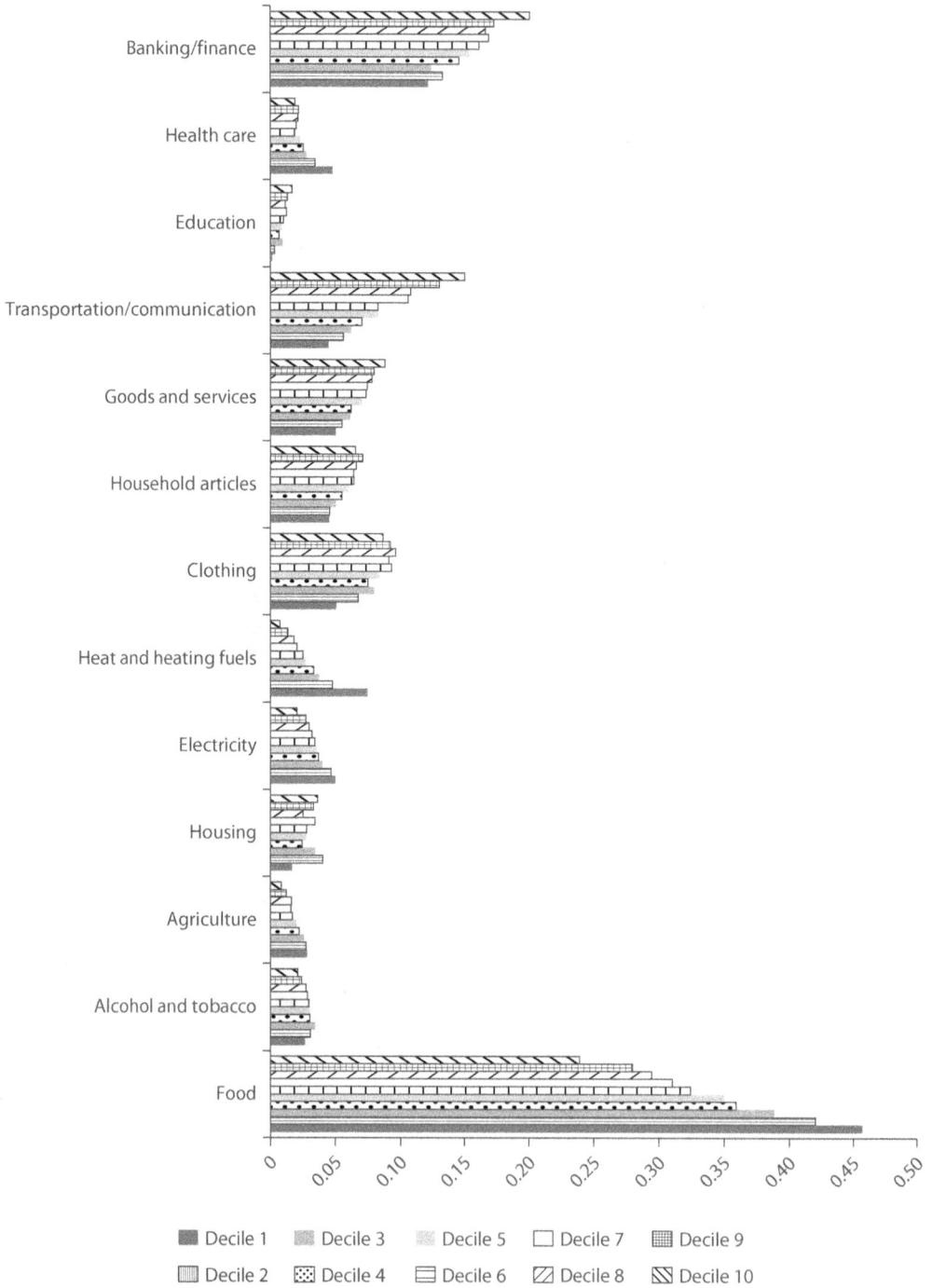

Source: Estimation based on HBS2009 and Belarusian Input–Output Table 2009.

beverages, the sector with the largest share of total expenditures for most households (figure 2.8), output prices increased by roughly 2 percent.

On the basis of the household consumption patterns shown in figure 2.8, figure 2.9 shows how the tax burden is distributed across different income groups. Every income group would see a cost increase on key consumption goods, such as food, clothing, and household articles. The extra expenditure in absolute terms is the highest for households in the top income decile, but the impact appears to be modestly regressive, as the percent of expenditures increased is slightly higher for lower-income households.

The direct and cross-subsidies have imposed rapidly increasing fiscal and quasi-fiscal costs. As a result of declining cost-recovery rates, both ZhKH and Belenergo incurred growing operational losses in the residential DH business. The total fiscal and quasi-fiscal costs, measured by the cumulative operating losses on residential DH services provided by both Belenergo and ZhKH, have risen from about 0.7 percent of gross domestic product (GDP) in 2005 to about 1.6 percent in 2012. Of this, ZhKH accounts for about 40 percent and Belenergo for the remainder. Figure 2.5 shows the cost of heat subsidies as a percentage of GDP.

Underpriced residential utility tariffs also create significant fiscal risks and macroecnomic vulnerabilities. Because Belarus continues to benefit from subsidized import prices for gas (less than half of the economic price imputed from the European gas price), current financial imbalances in the utility sector—while fiscally costly and harmful in terms of energy efficiency—have had a limited macroeconomic impact. However, the lack of tariff adjustments and low cost recovery of utility tariffs create significant risks. In case of price hikes for gas

Figure 2.9 Extra Expenditures from Imposing Implicit Tax on Industrial Consumers

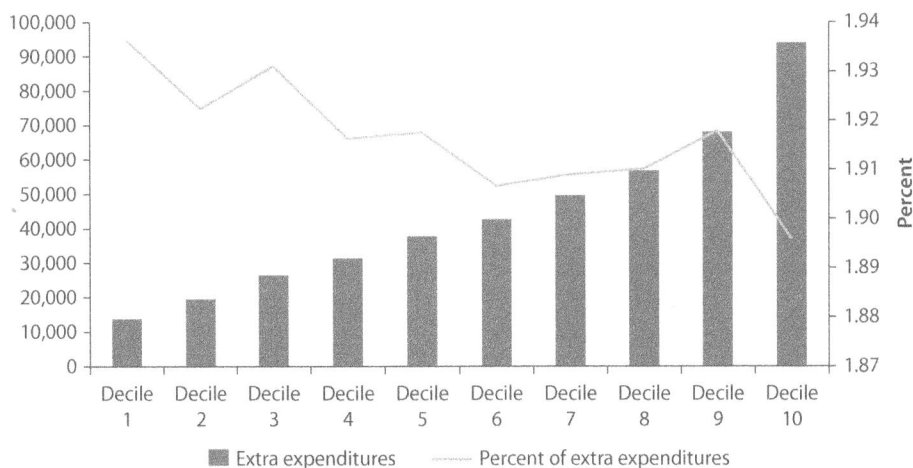

Source: Estimation based on HBS2009 and Belarusian Input–Output Table 2009.

Figure 2.10 Distribution of Heat Subsidies

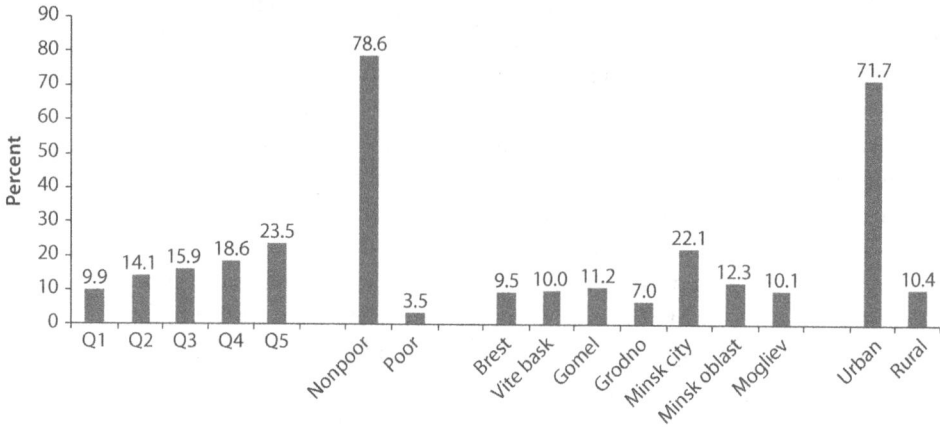

Source: Grainger, Zhang, and Schreiber 2015.

imports, the existing imbalances would amplify and likely induce fiscal and macroeconomic instability. At current tariff and consumption levels, financial losses in the district heating sector could more than double if Belarus were to import gas at market prices, imposing a significant fiscal and macroeconomic risk.

Although heat subsidies are expensive, they are poorly targeted and regressive; wealthy households benefit more than poor households from the subsidies. Rather than identifying the poorest households and allocating subsidies accordingly, they are given to all residential consumers regardless of income level. Because higher-income households tend to occupy larger living areas which require more energy for heating purposes, they receive a larger portion of overall heat subsidies: the top two income quintiles receive 42.1 percent of the overall heat subsidy, while the lowest two quintiles receive only 24 percent. Poor households and those in rural areas also receive a smaller portion of the heat subsidies: 3.5 percent to the poor and 10.4 percent to rural households. Figure 2.10 shows the distribution of heat subsidies.

Notes

1. The Council of Ministers makes a resolution on a quarterly basis establishing the subsidized residential heating and hot water tariff. The most recent, N 1166, set the tariff at 80,570 BYR per Gcal against an "economically justifiable" tariff of 300,000 BYR per Gcal. No information is publicly available on the methodology applied for determining the economically justifiable tariff, but it is thought to be a tariff that reflects the actual costs of heat production and delivery.

2. Resolution 220 mandates cross-subsidization between types of utility service and between customer classes to recover costs.

CHAPTER 3

What Is the Likely Impact of Tariff Reform?

Impact of the Reform

The impact of tariff reform depends on how the government of Belarus (GoB) decides to implement it. To be consistent with the government's vision to gradually phase out heat subsidies, this report explores three tariff increase scenarios in the medium to long term: one under a differentiated pricing regime (where customers of Belenergo and ZhKH are charged different tariffs) and two under a uniform pricing regime (where all customers are charged the same tariff). Table 3.1 describes the three tariff reform options and cost-recovery targets for 2015, 2017, and 2020. Because 61 percent of households, 81 percent in urban and 14 percent in rural areas, are connected to the district heating (DH) network in Belarus, a tariff increase will have a profound impact on many households. Specifically, increasing tariffs to 60 percent and full cost-recovery levels will significantly reduce the affordability of DH for households in the poorest quintile. At full cost-recovery levels under the uniform price scenario, 30 percent of households in the poorest quintile living in urban areas would spend more than 20 percent of their income on heating compared to 0.2 percent under the 2012 price. Under the differentiated price scenario, 12 percent of households in the poorest quintile living in rural areas would spend more than 15 percent of their income on heating. Figure 3.1 shows the impact of tariff increases under different price scenarios by income group.

Table 3.1 Tariff Reform Scenarios, 2015, 2017, and 2020

	2015		2017		2020	
	Cost recovery goal (%)	Pricing	Cost recovery goal (%)	Pricing	Cost recovery goal (%)	Pricing
Scenario 1	30	Uniform	60	Differentiated	100	Differentiated
Scenario 2	30	Uniform	60	Uniform	100	Uniform
Scenario 3	30	Uniform	45	Uniform	60	Uniform

The results are somewhat different for rural and urban customers depending on whether a differentiated or uniform pricing regime is adopted. Rural poor households in the poorest quintile that rely on DH services for heating are most vulnerable under the differentiated pricing scenario and are expected to spend 23 percent of their income on DH at full cost-recovery levels. This is because the cost of DH service provided by ZhKH is about double that of Belenergo's. ZhKH needs to raise tariffs to a much higher level in order to reach the targeted cost-recovery rate.

In contrast, the urban DH consumers in the poorest quintile will be most heavily affected under the uniform pricing scenario. They are expected to spend 21 percent of their incomes on DH services at full cost-recovery levels. On average, urban households spent more than rural households on DH, possibly

Figure 3.1 Financial Burden of District Heating on Households after Tariff Increases

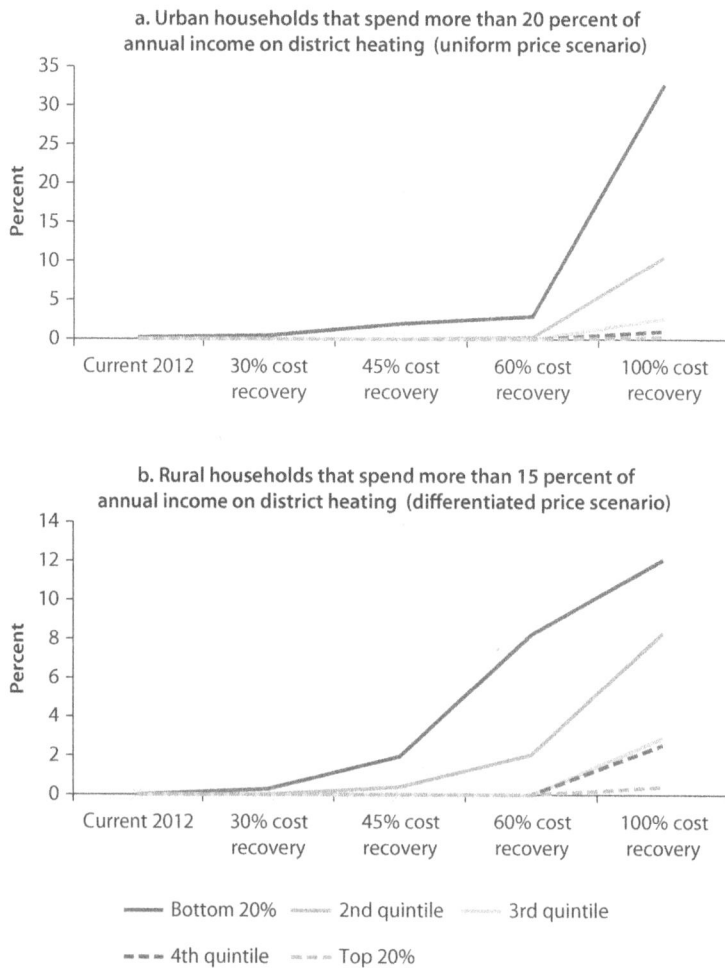

a. Urban households that spend more than 20 percent of
annual income on district heating (uniform price scenario)

b. Rural households that spend more than 15 percent of
annual income on district heating (differentiated price scenario)

Source: Simulation based on HBS 2012.

because rural households are more likely to use and have access to substitutes such as wood, peat, and coal. Figures 3.2 and 3.3 show affordability levels of DH services for households that are connected to DH under various pricing scenarios by income group and settlement type.

Figure 3.2 Budget Share of District Heating Expenditure under Uniform Pricing Regime

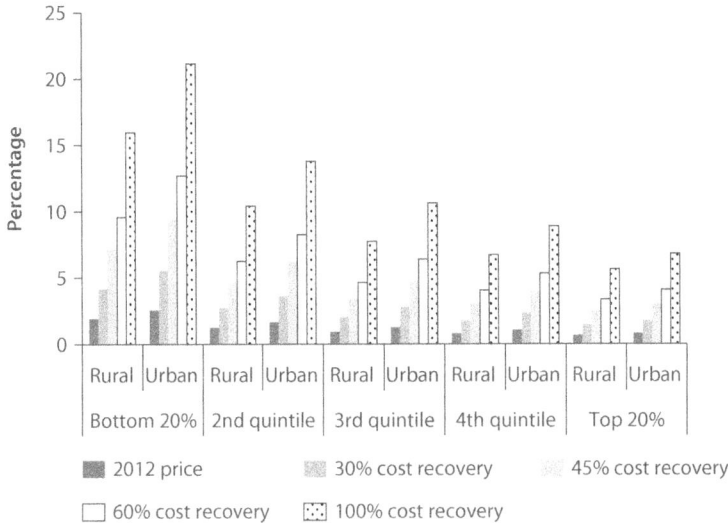

Source: Simulation based on HBS 2012.

Figure 3.3 Budget Share of District Heating Expenditure under Differentiated Pricing Regime

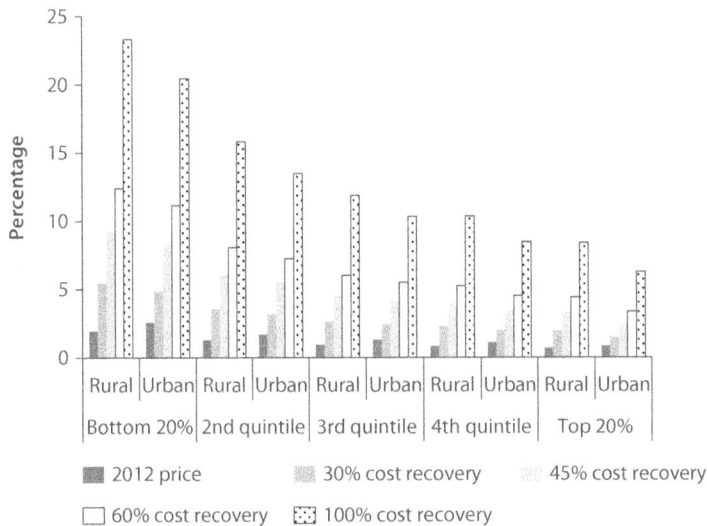

Source: Simulation based on Household Budget Survey 2012.
Notes: Data show the budget share of district heating of households that are connected to district heating services.

Belarus Heat Tariff Reform and Social Impact Mitigation
http://dx.doi.org/10.1596/978-1-4648-0696-4

The impact of higher heating tariffs will be most acutely felt in winter. The heating season in Belarus normally begins in October and ends in April. Accordingly, household expenditure on heat tariffs spikes in the first and last quarter of the year. In 2012, households that were connected to DH spent, on average, 1.5 percent of their monthly income on DH services in the fourth quarter and 4.5 percent in the first quarter. At full cost-recovery levels, urban households in the bottom 40 percent income group, under the uniform price scenario, will spend approximately 23 percent of income on DH services in the first quarter; rural households in the bottom 40 percent, under the differentiated price scenario, will spend 21 percent. Figure 3.4 shows the impact of tariff reform by price scenario, location, and time of year on households in the lowest two quintiles.

In addition to added heating expenses in winter months, fruits and vegetables are more expensive; and electricity bills also go up due to shorter daylight time.

Figure 3.4   Impacts of Tariff Increases during Q1 through Q4

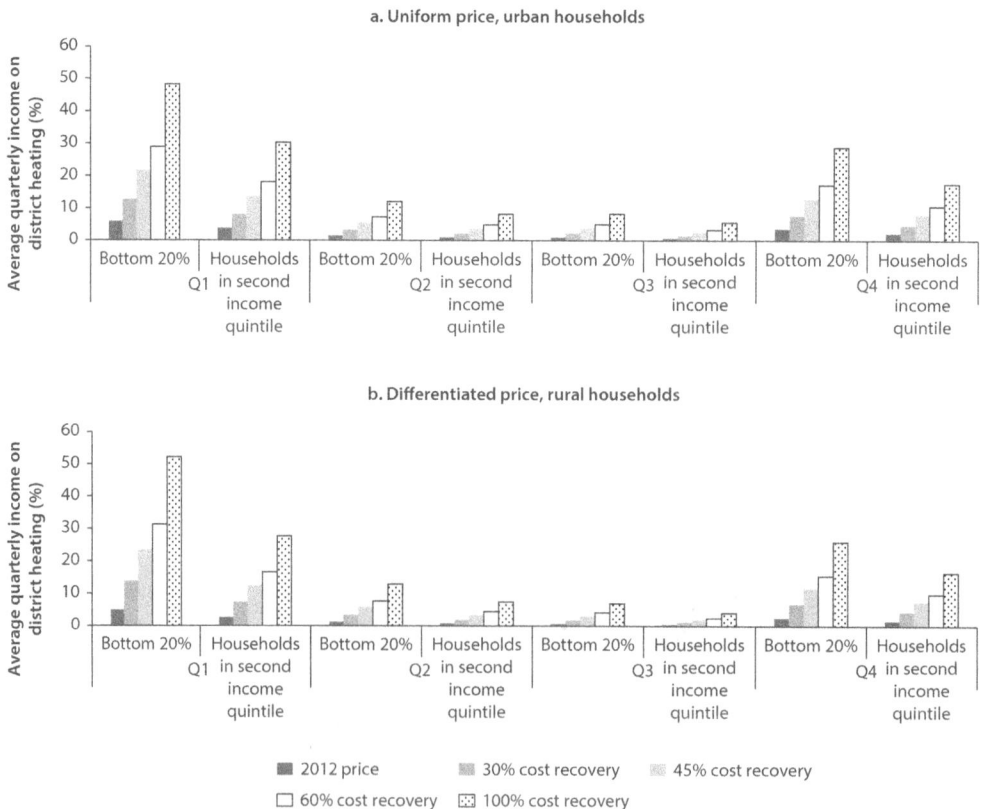

a. Uniform price, urban households

b. Differentiated price, rural households

■ 2012 price   ▦ 30% cost recovery   ▨ 45% cost recovery
□ 60% cost recovery   ▦ 100% cost recovery

Source: Simulation based on Household Budget Survey 2012.
Note: Q1–Q4 refers to quarters 1–4.

Penalties for delay in paying utility bills are more likely to occur in winter months. All of these have made the financial situation of low-income households most stressful during the winter season.

To cope with a heat tariff increase, poor households are more likely to cut back on consumption of food and clothes than on heat. In focus group discussions, 90 percent of the participants from poor households without social benefits, 67 percent of those from poor households with social benefits, and 53 percent of those from middle-income households said they would reduce food expenditures to cope with a tariff increase. Figure 3.5 describes various coping strategies indicated by focus group participants in case of an increase. Appendix B details the methodology and scope of focus group discussion. The majority of buildings (92 percent) in Belarus are not equipped with apartment-level meters and heat regulators. Participants in the focus group also complained about their inability to control heating costs. In case of overheating, householders prefer to open windows rather than report to service providers, in order to avoid conflict with neighbors

Tariff increases are expected to have a more pronounced impact on women than men as they are typically paid less and have fewer job employment opportunities. Moreover, single-parent families are more often headed by women,

Figure 3.5 Common Coping Strategies in Response to Tariff Increases

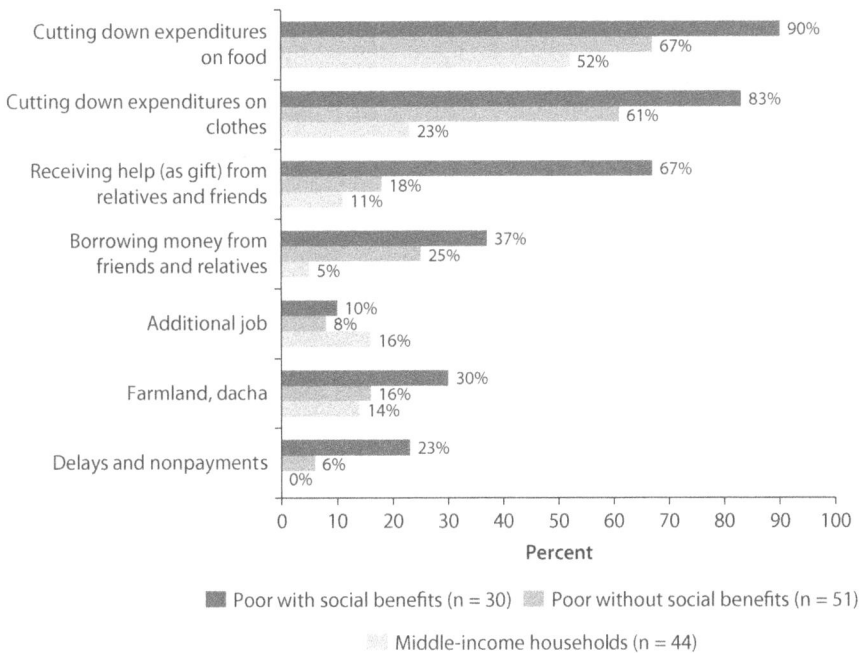

Legend: Poor with social benefits (n = 30)   Poor without social benefits (n = 51)   Middle-income households (n = 44)

including single-female pensioners, and women with young children often have difficulty finding part-time work. In the first quarter of 2012, 62 percent of women were employed compared to 69 percent of men. In 2011, women in Belarus earned 26.3 percent less than men.

In addition, focus group discussion reveals that there are noticeable gender differences in coping strategies. While men report they would work more or change jobs in response to higher utility payments, women are more likely to reduce their expenditures on food or to seek help from relatives and friends. Women are generally more sensitive about potential tariff increases and report strong insecurity about their future well-being in case of tariff increases.

On the positive side, the proposed tariff increase would generate large fiscal savings and utility revenues. It is expected that, with the increase in tariffs to the full cost-recovery level, the total fiscal savings will amount to 1.6 percent of gross domestic product (GDP) by 2020. Tariff reform under the differentiated pricing scenario will also make Belenergo heat sales profitable, allowing for more investments in new infrastructure and energy efficiency measures. Figure 3.6 shows the fiscal savings from each pricing scenario.

Lowering cross-subsidies would also open fiscal space to allow for reductions in electricity tariffs for nonresidential consumers. If nonresidential electricity prices are decreased to the level of cost of service, the average unit energy cost of manufacturing could decrease by about 24 percent. The wood, textile, food, and paper industries, which have some of the lowest export share in output

Figure 3.6 Fiscal Savings Generated from Different Tariff Increase Scenarios, 2015, 2017, and 2020

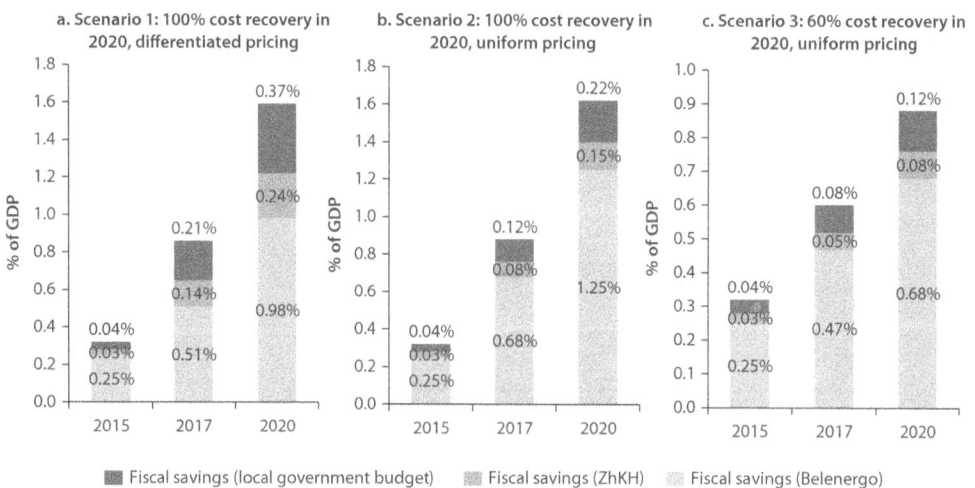

Figure 3.7 Share of Export in Total Output

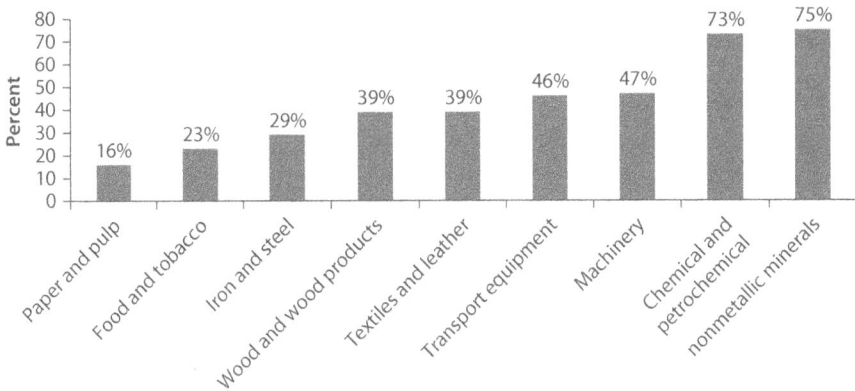

Source: Belarus Foreign Trade Database, 2009 data.
Note: The paper, food, wood, and textile sectors had some of the lowest export share.

(figure 3.7), would benefit the most because they use more energy to produce US$1 of manufacturing value added and use more electricity (rather than gas) for production than other sectors. The energy cost of manufacturing for the four industries would be reduced by between 25 and 28 percent after the removal of cross-subsidies.

CHAPTER 4

How Can Tariff Reform Be Best Implemented?

Implementation of the Reform

How can the proposed tariff reform be implemented so that it is socially acceptable and fiscally beneficial and does not impose undue hardship on poor and vulnerable households? We recommend the following three approaches:

1. Enhancing customer communications and engagement
2. Improving social protection mechanisms
3. Encouraging investments in supply-side and demand-side energy efficiency.

The subsequent paragraphs describe each of the approaches in more detail. The sequence of the reforms is important to its success. Promoting customer understanding and winning trust is a critical first step which can then run in parallel with the medium to long-term efforts to improve social protection mechanisms and encourage investment in energy efficiency.

Focus group discussion and stakeholder analysis reveal that there is limited support for tariff reform among residential consumers because of their lack of knowledge and awareness of tariff setting and reform processes. The lack of information has been identified as one of the most aggravating factors for residential consumers as they often learn about tariff increases only after receiving the bill.

Infrequent and insufficient interaction between district heating (DH) providers and the customer base has also contributed to consumers' lack of trust in service providers, which further undermines support for reform. The key to increasing public acceptance of DH tariff increases is developing a comprehensive communication and consumer engagement strategy that fosters a culture of inclusion, along with a public awareness campaign explaining the need for reform.

Focus group discussions reveal that the public is more likely to support a tariff increase if there are corresponding increases in wages and other benefits (such as pensions and social assistance), an upgrade in DH service (utilization of new technologies and modernization of equipment), greater transparency from service providers reflected in clearer bills, and introduction of individual metering and heat controls.

Communication strategies should address the public's concerns by explaining the rationale for a tariff increase, the government's plans for protecting the poor, and the means by which it will improve the transparency and accountability of the DH sector. Similar efforts in the neighboring former Soviet Union countries present useful examples of a reform strategy in which communication was the key to reform implementation (appendix C). Communication efforts should also focus on motivating people to save energy and gaining public support for the development and upgrading of the energy system.

Communication activities should involve all key government institutions (at both national and local levels) and utilities in a systematic manner and target consumers at the regional, city, district, and village levels. This would require a national agency to coordinate various ministries and agencies to develop guidelines on the best methods of interacting with different consumer groups. Messages from various organizations should be consistent and complementary, rather than contradictory. The national agency could offer such ready-made messages to communicate with the public.

Public information products, such as social advertisements, posters, leaflets, and brochures should be developed at the national level. This is more cost-effective than doing it locally and ensures that the public hears a "single voice." The consumers' most preferred channels of communication for information on utility services are utility bills, national and local mass-media, tenants meetings, hotlines, information boards, and Internet.

Residential consumers' engagement in policy debates and decision making could also contribute to a culture of shared responsibility in the governance process while providing useful feedback on the reform. Such measures would make people feel they are owners of the reform and share responsibility for its outcome. Because public sentiment reflects a sense of exclusion from the policy debate, various strategies could be applied to address this issue. For example, people could be invited to comment on the reform process through feedback mechanisms, such as public surveys and online consultations. Institutional arrangements could be made to ensure citizen's feedback is processed, analyzed, and factored into decision making.

Communication campaigns and consumer engagement efforts should particularly target women as focus group discussions reveal that women are more involved than men in managing utility bills and interacting with service providers regarding the quality of services.

Providing more and better information and feedback mechanisms would also help improve the transparence and accountability of utilities and address consumers' key concerns. In addition, consumer monitoring mechanisms could be established by introducing performance benchmarks and key performance indicators, such as fuel efficiency and the cost of production.

The existing social protection mechanisms in Belarus are not well suited to protecting poor and vulnerable groups if tariff reforms are implemented. This is because most of the social protection benefits are categorical, poorly targeted, and inadequate. The only income-tested program is the Public-Targeted Social Assistance Program (GASP), which provides short-term (six months of the year) income support to financially distressed families. It remains tiny: In 2012, its maximum coverage was 2.8 percent of population; its budget was 0.08 percent of gross domestic product (GDP). The government has considered reintroducing the Housing and Utility (H&U) subsidies program that existed until 2009. The program provides social assistance to households whose housing and utility expenses exceed 15 of their income (in rural areas) and 20 percent (in urban areas). However, the scheme does not account for income disparities and therefore is not sufficient to protect the poor.

There are two ways in which the targeting and coverage of social protection mechanisms could be improved to mitigate the adverse impact of tariff increases on the poorest 20 percent of the population:

1. Expanding or topping-up GASP
2. Refining the H&U program.

In addition, introducing levelized payment plans presents another cost-effective strategy to help households manage the seasonality of utility expenses.

The first option to mitigate the impact of tariff reform is to expand or restructure (top-up) the existing poverty-targeted cash transfers program, GASP. To reach the target coverage of the poorest 20 percent, GASP would need to be expanded 10 times, and the income threshold greatly raised to allow more households to qualify for the benefit. As a result, the program budget would considerably increase to reach as much as 1 percent of GDP in 2015 and stay high in the following years. On the other hand, the program would yield a high reduction in poverty: The poverty rate among the poorest 10 percent would drop twofold after the transfer (table 4.1).

To reduce the program's cost, GASP may be augmented by a supplemental flat benefit. The supplement would be paid on top of the regular GASP benefit received by those who pass the income threshold. Such a flat benefit could also be offered to other households affected by the tariff increase with incomes above the regular GASP threshold but below a separate threshold established specifically for the top-up.

Given the low coverage of GASP, however, such a supplemental benefit would significantly exceed the "base" benefit program and would look more like a separate program. In addition, the current restrictive eligibility criteria in

Table 4.1 Benefit Coverage, Targeting Accuracy, and Fiscal Cost of GASP and H&U Benefits

	Benefit coverage		Targeting accuracy		Budget per year, % GDP	
	2015	2017	2015	2017	2015	2017
Expand GASP (20% of population)						
1st decile	52	51	42	41	0.43	0.36
2nd decile	48	52	21	24	0.22	0.22
3rd–10th deciles	12	12	37	35	0.38	0.31
Total	18	19	100	100	1.03	0.89
Expand GASP (10% of population) +						
Top up GASP (10% of population)						
1st decile	100	100	59	59	0.26	0.25
2nd decile	81	83	20	23	0.09	0.1
3rd–10th deciles	2	2	21	18	0.09	0.08
Total	20	20	100	100	0.44	0.43
Old H&U benefit						
1st decile	5	21	48	25	0.002	0.01
2nd decile	1	10	15	12	0.001	0.01
3rd–10th deciles	1	5	37	63	0.002	0.03
Total	1	7	100	100	0.005	0.05
Refined H&U benefit						
1st decile	27	61	84	60	0.012	0.04
2nd decile	3	18	12	16	0.002	0.01
3rd–10th deciles	0	3	5	25	0.001	0.02
Total	3	10	100	100	0.014	0.07

Source: Simulation based on HBS 2012.

GASP may impede households in the second decile from getting the top-up benefit, even with a higher-income threshold. Therefore, it would be important to combine the supplemental benefit with some expansion of GASP, including reconsidering GASP's ability to be more inclusive without compromising on the targeting accuracy of the benefits.

At lower cost, the GASP top-up option would have a poverty impact comparable to its expansion scenario and would be more efficient. Table 4.2 summarizes the simulation results under a scenario when the regular GASP benefit is expanded to cover the first decile, while the supplemental flat energy benefit is provided to the population in the second decile. This way, the entire program ensures the target coverage of the bottom 20 percent population but incurs half the cost (0.5 percent of GDP) of the GASP expansion scenario. It achieves a similar poverty reduction impact with better targeting accuracy.

Several other considerations should be taken into account for GASP adjustment. First, the eligibility rules would need to be revisited to allow quick response and inclusion of more households. Second, according to stakeholder analysis and focus group discussions, the GASP program is

Table 4.2 Poverty Impact of GASP and H&U Benefits

	Total population		1st decile		2nd decile	
	2015	2017	2015	2017	2015	2017
National poverty line						
Before transfers	1.8	4.3	18.2	42.4	0	0
Expand GASP	0.9	2.1	8.6	20.6	0	0
Top-up GASP	0.8	1.6	8.02	15.91	0	0
Old H&U benefit	1.8	4.1	18.1	41.4	0	0
Refined H&U benefit	1.8	3.9	17.5	38.9	0	0
H&U poverty rate (H&U costs greater						
than 15% of total incomes per year)						
Before transfers	1.1	5.9	5.3	18.9	1.4	7.7
Expand GASP	0.8	4.1	3.1	8.3	0.9	3.8
Top-up GASP	0.5	3.5	0.8	2.5	0.6	1.9
Old H&U benefit	1	4.9	4.9	16	1.4	6.1
Refined H&U benefit	0.5	3.5	0.7	1.3	0.5	3.9

Source: Simulation based on HBS 2012.

Note: National poverty line in November 2012: BYR 880030 per capita per month. Welfare indicator: Total income per capita.

associated with the stigma of poverty. The program would need to be rebranded to improve its image. Third, the current six-month payment period of GASP limits the program's efficiency. The payment period for benefits and/or top-ups should be 12 months for the time of the tariff reform or be specifically linked to the heating season. Fourth, administration of the benefits is fully handled at the local level with local authorities determining the eligibility and paying the benefits from their budget. This means that the budget pressure on poorer regions would be higher as the demand for benefits would be higher there. Furthermore, in their combined roles of checker and payer, the local authorities could restrict access to the benefits to offset increased spending. To mitigate this risk, the central budget may guarantee additional transfer for the regions in case the demand exceeds available budget resources. Finally, the oversight and control functions should be strengthened to mitigate the risk of system abuse by both providers and beneficiaries.

The H&U subsidy program was phased out in 2009. Some advantages to reintroducing the H&U subsidy include public familiarity with the program and its specificity to energy tariffs. Furthermore, if reintroduced, the government of Belarus (GoB) will be able to learn from past lessons to create a more efficient and streamlined program. However, the design of the program does not differentiate consumption patterns and income levels of households. Therefore, an important drawback of this program is that it does not provide any support to a large share of genuinely poor households that use alternative fuels for heating or which are too poor to afford to spend 15 or 20 percent of their income on heating. Empirical evidence suggests that this type of programs is also quite regressive, with the majority of the benefits going to middle- and upper-income households.

To improve the coverage of the poor with H&U benefits, households may be compensated for a share of the heating bill based on their per capita income. Such a refined formula would use an income test to determine eligibility and to differentiate benefit payments based on income levels. For example:

- Households from the first and second deciles would be compensated for the expense above 10 and 15 percent of their income, respectively
- Households from the second and third quintiles would be compensated for the expense above 20 percent of their income.

The simulation results presented in table 4.2 suggest that the refined H&U benefit would cover more households in the lowest deciles as compared to the "old" benefit (84 and 60 percent vs. 48 and 25 percent in 2015 and 2017, respectively) and fewer funds would be leaked to better-off households (5 and 25 percent vs. 37 and 63 percent of benefits going beyond the bottom quintile in 2015 and 2017, respectively). Spending for the refined benefit would be higher in 2015 but still remain at a sustainable level of 0.014 percent of GDP; in 2017, the two programs would require a similar budget.

The refined H&U benefit would significantly reduce the incidence of households with high H&U-related costs (H&U expenses above 15 percent of annual income) among those belonging to the bottom decile (from 18.9 to 1.3 percent in 2017). In comparison, the "old" H&U benefit would only reduce the percentage of households in the bottom decile which spent more than 15 percent of annual income on H&U from 18.9 to 16 percent in 2017 (Table 4.1). Table 4.2 compares the old H&U subsidy program to the refined program in terms of benefit coverage, targeting accuracy and fiscal cost for 2015 and 2017.

Although it is more convenient for households to receive the in-kind benefit (no need to collect cash transfers every month), H&U benefits are less transparent than GASP and distort incentives for energy efficiency once household expenditures are above the 15/20 percent threshold. Table 4.3 compares the advantages and disadvantages of the GASP and H&U programs.

The H&U benefits could be further streamlined if administered under the Ministry of Labor and Social Protection rather than the Ministry of H&U. It is not the role of H&U service to determine vulnerability and thus eligibility for social assistance. The Ministry of H&U could help social protection units to calculate/verify H&U costs and to check eligibility and determine the benefit. In fact, relatively simple software may offer a technological solution to checking eligibility and determining the benefit size.

Table 4.3 Advantages and Disadvantages of GASP and H&U Benefits

	Advantages	Disadvantages
Expand GASP	• High poverty impact: targeted to the poor by design (means-test)	• Dramatic increase in program costs
	• Implementation infrastructure is in place	• System administrators may limit entry and hence budget expenses
		• Social stigma to accept assistance from the program
Top-up GASP	• Less costly than GASP expansion	• Given the very low coverage of GASP, top-up would need to be combined with some expansion of GASP
"Old" H&U benefit	• Relevant to energy consumption by design	• Less transparent
	• Providing larger coverage	• Weak targeting accuracy and possible leakage to better-off households
		• Lower poverty impact
		• Cumbersome administration outside of social protection system
		• Distorting incentives for energy efficiency
Redefined H&U benefit	• Stronger poverty impacts	• Less transparent
	• More accurate targeting	• Cumbersome administration outside of social protection system
		• Distorting incentives for energy efficiency

That would also reduce the workload for the social workers. Applying for income support in a single unit makes even more sense, given the similarity of information/data collected for GASP and H&U benefits. It would also provide a more client-centered approach, allowing clients to apply for any benefit in a single place.

Administration of both GASP and H&U benefits is currently fully handled at the local level with local authorities determining the eligibility and paying the benefits from their budget. This means that the budget of poorer regions would be put under higher pressure because of higher demand for benefits. The literature suggests that centrally financed social protection programs generally work better, while locally financed ones typically suffer from low coverage, payment arrears, and poor protection offered in poor localities. For the above-discussed mitigation measures to work, it is recommended that the financing arrangements be switched from local to central financing, with complementary investments in a stronger oversight and control system.

Customers' energy bills are the highest in winter months and lowest in summer months. This can be hard for households to manage from a cash-flow perspective. Under a levelized payment plan, customers make identical, fixed payments every month, regardless of their actual consumption in any given month. Levelized payment plans help ease the cost of heating during the coldest months and recoup utility revenues during the summer months. A levelized payment plan would average annual energy consumption over a 12-month period, basing monthly charges on past energy usage and estimated future rates. Bills may be adjusted every six months to minimize the difference between actual costs and plan amounts.

Investments in energy efficiency can further reduce consumer bills, and they offer long-term recurrent assistance to energy affordability. Despite recent achievements in energy efficiency improvement in Belarus, there are, nevertheless, more improvements that can be made on both the supply- and demand sides in the DH sector. It is important to note that many of the measures described below have a cost. A tariff increase will therefore be needed to fund many of the measures.

The heat losses of both Belenergo and ZhKH have consistently declined; however, the cost of heat supply still varies widely among the oblasts, indicating opportunities for further improvement. For example, in 2012, Gomel oblast had the lowest cost at US$55.2/Gcal, which is less than half of that in the Grodono Oblast at US$117.3/Gcal. The difference between the lowest and highest cost of heat supply within one oblast is also substantial and ranges from 24 to 51 percent (figure 4.1).

On the demand side, 84 percent of the residential buildings in Belarus were built before 1993 based on Soviet norms (figure 4.2). The average heat energy consumption of these buildings (for both heating and hot water) is around 230 kilowatt hour per meters squared per year, almost twice as much as new building stocks developed under new thermal insulation standards and energy-efficient engineering systems (figure 4.3). Retrofitting and upgrading old building stocks therefore presents potentially large energy savings.

Figure 4.1   Substantial Variaton of Heat Production Cost among Oblasts

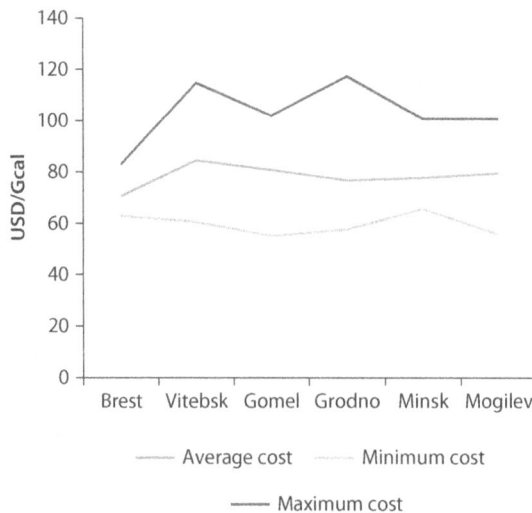

Source: Ministry of Housing and Utilities.

Figure 4.2   Distribution of Housing Stocks, by Heat Consumption, 1995–2012

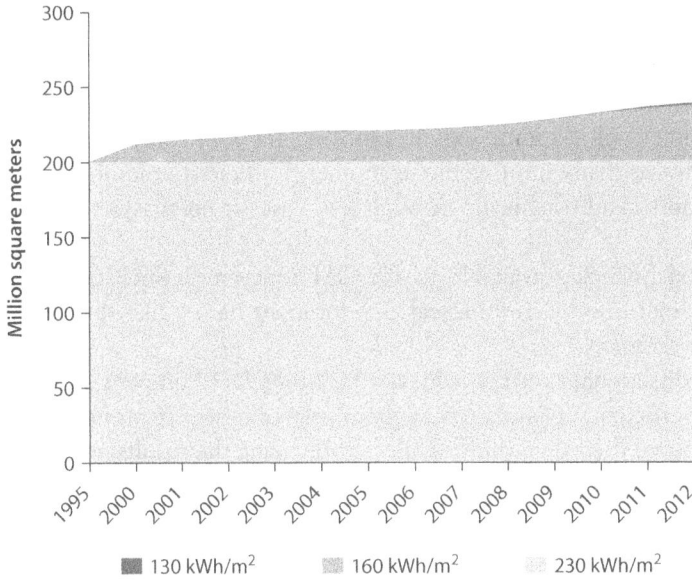

Figure 4.3   Heat and Hot Water Consumption, by Building Type

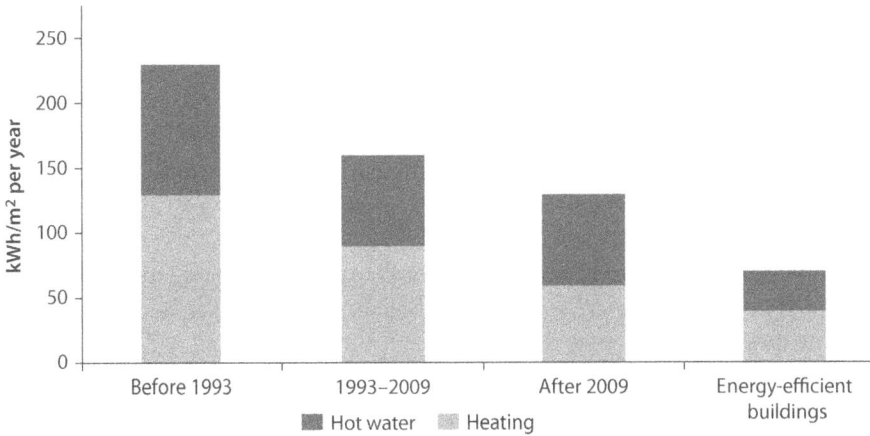

Source: Ministry of Housing and Utilities.

Recommendations to improve the energy-efficiency incentives on the supply side are as follows:

- Improve the incentives of the DH companies to increase their production efficiency. A pilot project launched by Brest Oblast in January 2014 provides an example of how to improve such incentives. The pilot foresees that all savings (measured by actual results as compared with an initial plan) achieved by the company at the end of the year can be kept by the DH companies rather than returned to local government. The savings may be used for additional investments or for financial reward of the personnel who contributed to the improvement in energy efficiency.

- Gradually reduce the subsidies and cross-subsidies related to heat production. Reducing subsidies will, to some extent, motivate the DH companies to reduce costs. Whereas many improvements in energy efficiency require capital expenditure, some can be achieved through low-cost (or no-cost) changes in management.
- Make additional financing available to the DH companies. Such financing could include financing by multilateral development banks, like the World Bank, or by government.
- Publicize and disseminate results achieved by other DH companies in improving energy efficiency. Disseminating the results of other efforts helps DH companies improve their own performance. Publicizing the results also helps owners and customers understand how their DH service provider compares in performance to others.

Based on the case studies of three representative towns and analysis of technology gaps in western DH systems, the following supply-side energy-efficiency measures are recommended to improve production efficiency. The investments also contribute to improving the efficiency, quality, and sustainability of urban utility network and infrastructure upgrading.

- Replacing old, low efficiency boilers with newer, more efficient ones. Many DH plants and boilers in Belarus are in need of rehabilitation. Some boilers have been running for more than 30 years and are past their technical life span. As a result, these boilers are running at 50–60 percent efficiency levels.
- Converting from natural gas boilers to boilers using renewable fuels. Heat generated in state-of-the-art wood-fueled boilers is estimated to be about 10 percent more cost effective than traditional natural gas-fired boilers at today's prices. Since natural gas import prices are expected to increase more than the price of wood, the cost efficiency of wood-fueled boilers is likely to improve over time.
- More rational boiler sizing. Some boilers in Belarus run at only 30–40 percent of their design capacity, especially during the summer months when typically only hot water service is needed. This results in higher fuel use because boilers running at lower-capacity factors typically are also less efficient. Such inefficiency could be avoided if more smaller-capacity boilers are installed. Some of these units could therefore be shut down completely during the summer months, leaving a smaller number of units to run at relatively high-capacity factors. The units shut down in summer could be restarted in winter to meet heating demand.

- Replacing steam with hot water boilers. In some areas, steam boilers are still used, a legacy of times when there was a larger industrial load. These steam boilers are now used primarily to serve residential customers (for example, in Starye Dorogi). Steam boilers are less efficient in serving the needs of residential customers than hot water boilers.
- Replacing networks. Replacement of worn-out network parts, where the heat losses and water losses are high, with preinsulated pipes is one of the most important investment priorities. Payback periods for networks are usually longer than paybacks for other DH investments, but investment in the network is essential to keeping the DH system operating in a sustainable way.
- Reduction of the network dimension and optimization of the network routes. Due to the reduction in heat loads over (last decades), the routing and dimensions of some network parts need to be changed. As in other transition countries, consumption has dropped as some consumers have left the DH system and energy efficiency measures have been introduced. In such cases, the dimension of the DH pipes should be reduced and a more direct route from the heat generation plant to the consumers considered in order reducing heat losses in networks. For instance, in Volkovysk one consumer with a capacity of 0.4 Gcal is connected with the DH through a 1.5 kilometer network (one pipe counting), without any other consumers connected to this network branch.
- Replacing pumps. Replacement of old, low-efficiency network pumps, which are often oversized, has a big potential to reduce electricity costs.

The economic rates of return of the above investments depend on site-specific details, such as the efficiency of old boilers and the number of operating hours per year. Table 4.4 describes the economic performance of some of the recommended investment components based on the data of the case study towns.

Network renovation projects are usually high-cost investments with long payback periods. The investment cost per kilometer and achieved savings depends very much on the diameters of the installed network parts. However, replacement of the network is often a technical must in order to keep the whole DH system operational. Table 4.5 shows results of the economic analysis of network renovation programs in the towns studied.

Table 4.4 Economic Analysis of Supply-Side Energy-Efficiency Measures in Case Study Towns

Boiler investment	Investment cost (US$ million)	Efficiency of old boilers (%)	Efficiency of new boilers (%)	Total capacity (MWh)	Annual heat production (Gcal)	Reduction of gas use ('000 m³)	ERR (%)	NPV (US$ million)
Replacement of base load NG boiler	522	85	94	9	54,990	569	49	1.00
Replacement of peak load NG boiler	522	85	94	9	11,526	119	4	−0.17
Replacement of NG boilers to wood-chip boilers	8,520	n/a	84	19.5	38,088	5,303	13	1.49

Belarus Heat Tariff Reform and Social Impact Mitigation
http://dx.doi.org/10.1596/978-1-4648-0696-4

Table 4.5 Economic Analysis of Network Renovation in Case Study Towns

Investment	Investment cost (US$ million)	Length	Losses before project (%)	Losses after project (%)	Gas saving ('000 m³)	ERR (%)	NPV (US$ million)
Replacement of old pipes with PI pipes	0.82	1.77	19	16.3	333	12.60	0.12
Replacement of old pipes with PI pipes	3.00	8.5	8.5	5.6	490	0.30	−1.46

Note: PI = preliminary insulated.

Demand-side energy-efficiency measures reduce household energy consumption and expenditures, which in turn allow service providers to reduce production and, in the long run, capital investment. Demand-side measures can also save customers money, limiting the impact of tariff increases on monthly bills. Demand-side measures that the GoB can put in place include the following:

- Changing from central district heating substations to individual substations. Replacing central district heating substations with individual substations can lead to substantial savings. With central district heating substations, temperature is controlled at the central district heating substation, which provides heat to a group of buildings. The supply of heat to each single building depends on the average demand of the buildings connected to the central district heating substation. With building-level individual substations, temperature is controlled independently in each building, and the heat supplied to the building depends on the actual consumption of each.
- Building thermal renovation. Most buildings in Belarus were constructed according to Soviet norms, and heat losses though the building construction elements are high. The highest potential for energy savings lies in insulating external walls, replacing windows, and insulating roofs.
- Apartment-level heat metering and regulation. Heat is currently metered at the building level, and residents do not have the ability to measure and control heat consumption in their flats. Internal piping in most buildings is vertical, making flat-level heat metering complicated since all customers take heat from the same pipes. It is assumed that, in parallel with increases in heating tariffs, the incentives to have flat-level heat regulation, metering, and billing will increase.

At the current level of the residential heat tariff, however, none of the suggested energy-efficiency measures would be feasible. As shown in table 4.6, payback periods of all investment components are very long, with the payback of the whole investment more than 75 years.

As shown in table 4.7, by increasing tariffs to reflect cost, installation of thermostatic valves and individual substations would become most profitable. The net present values (NPVs) of these investment components are positive, starting from 2017, under all price increase scenarios. Under a full cost-recovery

Table 4.6 Economic Analysis of Demand-Side Energy-Efficiency Measures under Current Tariff

Energy-efficiency measures	Unit	Unit cost (USD)	Total investment in a typical building (USD)	Simple payback period	IRR	NPV	Energy savings potential (Gcal)
Windows replacement							185.6
Double pane windows	m²	100	62,480	70.5	−0.106	−49,253	
Triple pane windows	m²	150	93,720	73.2	−0.108	−74,298	
External wall	m²	65	157,625	106.7	−0.133	−130,717	214.1
Roof insulation	m²	30	31,170	105.5	−0.132	−25,821	42.8
Radiator thermostatic valves	piece	40	7,176	29.1	−0.037	−4,427	35.7
House-level heat substation (ITP)	piece	15,000	15,000	20.3	−0.002	−7,347	30
Total investment			367,171	75.4	−0.11	−242,610	508.2

Source: Estimation based on data of an average building in the case study towns.
Note: ITP = individual substation.

Table 4.7 Economic Analysis of Demand-Side Energy-Efficiency Measures under Different Tariff Increase Scenarios

Energy-efficiency measures	2015			2017			2020		
	Simple payback period	IRR (%)	NPV	Simple payback period	IRR (%)	NPV	Simple payback period	IRR (%)	NPV
Scenario 1, Belenegro									
Windows replacement									
Double-pane windows	29	−3	−38,175	20	0	−30,742	12	6	−12,921
Triple-pane windows	30	−4	−58,297	21	0	−47,171	13	6	−21,819
External wall	43	−7	−112,253	31	−4	−99,416	18	1	−70,163
Roof insulation	43	−7	−22,128	30	−4	19,561	18	1	−13,710
Radiator thermostatic valves	12	6	−1,350	8	12	790	5	25	5,665
House-level heat substation (ITP)	8	12	1,885	6	20	8,303	3	40	22,930
Total investment	31	−4	−192,143	22	−1	−157,055	12	5	−77,098
Scenario 1, ZhKH									
Windows replacement									
Double-pane windows	29	−3	−38,175	11	7	−8,626	7	17	23,490
Triple-pane windows	30	−4	−58,297	11	7	−15,616	7	16	30,774
External wall	43	−7	−112,253	17	2	−63,006	10	9	−9,479
Roof insulation	43	−7	−22,128	17	2	−12,278	10	9	−1,573
Radiator thermostatic valves	12	6	−1,350	5	28	6,858	3	57	15,779

table continues next page

Table 4.7 Economic Analysis of Demand-Side Energy-Efficiency Measures under Different Tariff Increase Scenarios *(continued)*

House-level heat substation (ITP)	8	12	1,885	3	46	26,508	2	110	53,272
Total investment	31	−4	−192,143	12	6	−57,535	7	15	88,772
Scenario 2									
Windows replacement									
Double-pane windows	29	−3	−38,175	14	4	−19,549	9	12	5,284
Triple-pane windows	30	−4	−58,297	15	3	−31,394	9	11	4,477
External wall	43	−7	−112,253	22	−1	−81,211	13	5	−39,821
Roof insulation	43	−7	−22,128	21	−1	−15,919	13	5	−7,642
Radiator thermostatic valves	12	6	−1,350	6	20	3,824	4	39	10,722
House-level heat substation (ITP)	8	12	1,885	4	32	17,406	2	68	38,101
Total investment	31	−4	−192,143	15	3	−107,295	9	10	5,837
Scenario 3									
Windows replacement									
Double-pane windows	29	−3	−38,175	19	1	−28,862	14	4	−19,549
Triple-pane windows	30	−4	−58,297	20	0	−44,845	15	3	−31,394
External wall	43	−7	−112,253	29	−4	−96,732	22	−1	−81,211
Roof insulation	43	−7	−22,128	28	−3	−19,024	21	−1	−15,919
Radiator thermostatic valves	12	6	−1,350	8	13	1,237	6	20	3,824
House-level heat substation (ITP)	8	12	1,885	5	22	9,645	4	32	17,406
Total investment	31	−4	−192,143	20	0	−149,719	15	3	−107,295

Source: Estimation based on data of an average building in the case study towns.
Note: ITP = individual substation.

tariff, the simple payback of these investments is only two to five years. The whole investment package also becomes economically feasible under 100 percent cost-recovery scenarios.

If supply- and demand-side energy-efficiency measures are implemented, households would spend, on average, 41–46 percent less than they currently do for heat energy. Supply-side measures account for about 9 percent, while demand-side measures contribute an additional 35 percent of energy savings. Table 4.8 outlines the average annual household expenditures on heat energy and projected savings from the implementation of energy-efficiency measures under different tariff increase scenarios.

Energy efficiency assistance in the form of preferential loans and grants to low-income households can be an effective policy response to improve energy affordability. One such example is the US Weatherization Assistance Program. The program, which provides low-income households with weatherization services, initially targeted heating (insulation and heating systems) but has been broadened over time to include cooling, appliances, and lighting. Eligibility for the program is mainly based on income levels, using thresholds defined according to the national poverty guidelines. The weatherization services are managed by local agencies and include a visit by an energy auditor, installation of the chosen energy-saving measures, and finally verification of the work by an inspector. A recent cost-benefit analysis suggests that for every US$1 invested under the program, US$1.80 is returned in reduced energy bills and US$0.71 is returned to ratepayers, households, and communities through increased local employment, reduced uncollectible utility bills, improved housing quality, and better health and safety.

The expected yearly fiscal savings from tariff increases are approximately US$ 0.41 billion. These fiscal savings can be allocated to fund social protection programs

Table 4.8 Average Annual Household Savings after Implementation of Energy-Efficiency Measures, 2015, 2017, and 2020

			Before energy-efficiency measures			After supply-side energy-efficiency measures			After supply- and demand-side energy-efficiency measures		
	Unit		2015	2017	2020	2015	2017	2020	2015	2017	2020
Heat consumption of average household	Gcal/y		9.2	9.2	9.2	9.2	9.2	9.2	5.9	5.9	5.9
Heating cost of average household	USD	Scenario 1 (Belenergo)	156	220	367	142	200	334	92	130	217
		Scenario 1 (ZhkH)	156	403	672	130	336	559	84	218	364
		Scenario 2	156	312	519	136	272	453	88	177	294
		Scenario 3	156	234	312	136	204	272	88	132	177
Reduction	Percent	Scenario 1 (Belenergo)				9	9	9	41	41	41
		Scenario 1 (ZhkH)				17	17	17	46	46	46
		Scenario 2				13	13	13	43	43	43
		Scenario 3				13	13	13	43	43	43

Note: Assumptions used in the estimates are as follows: Heat production cost (present): 65.4 USD/Gcal (weighted average of Baranovichi, Volkovysk, and Starye Dorogi); average size of household: 65 m².

and energy-efficiency investments to mitigate the impact of tariff increases on the poorest households. Fiscal savings can also be used to reduce commercial electricity tariffs to improve business competitiveness. The results in Table 4.9 show that when an effective social assistance package, together with the removal of heat subsidies, is properly designed and sequenced, it is possible for government to achieve positive fiscal gains while protecting the poorest households.

The aforementioned analysis shows that the burden of higher DH tariffs falls most heavily on low-income groups. However, the negative social impact is manageable if the government improves existing social protection systems and scales up energy efficiency programs. It should be recognized that, while a number of easy-to-achieve opportunities might be available, addressing energy efficiency comprehensively also requires longer term investments and it takes time for the benefits to reach the households. An effective social assistance package should therefore consist of both welfare transfers that offer immediate relief and an energy efficiency program that provides a sustainable long-term solution.

As noted earlier, the sequence of the reform is critical. The reform program will be most effective if the GoB does the following:

1. Leads with a consumer communication campaign
2. Puts in place better social protection mechanisms, including grants to demand-side energy efficiency targeting at low-income households
3. Takes measures to encourage investments in supply- and demand-side energy efficiency.

When mitigation measures are properly coordinated, the heat tariff reform will become more socially acceptable, consumers will benefit from better quality of services, and the government will achieve positive fiscal savings. Figure 4.4 illustrates a proposed timeline for implementing the package of reforms described in this note.

Table 4.9 Reform Packages with Positive Fiscal Savings

	Fiscal savings (US$ bln)					Energy-effiency grant (US$ bln)	Industry rebate (US$ bln)	Net fiscal savings (US$ bln)
Year	Total	Local budget	Industry cross-subsidies	Budget of social protection (US$ bln)				
2015	0.15	0.02	0.13	Refined H&U	0.01	0.12		0.02
2016	0.15	0.02	0.13	Refined H&U	0.01	0.12		0.02
2017	0.29~0.41	0.04~0.1	0.25~0.31	Refined H&U + Expand GASP	0.30			0~0.11
				Refined H&U + Expand + Top up GASP	0.19			0.09~0.21
2020	0.42~0.76	0.06~0.18	0.37~0.59				0.37~0.59	0.06~0.18

Note: Fiscal savings in 2017 and 2020 reflect the range under three tariff increase scenarios.

Belarus Heat Tariff Reform and Social Impact Mitigation
http://dx.doi.org/10.1596/978-1-4648-0696-4

Figure 4.4 Recommended Road Map for Implementing Reforms, 2015, 2017, and 2020

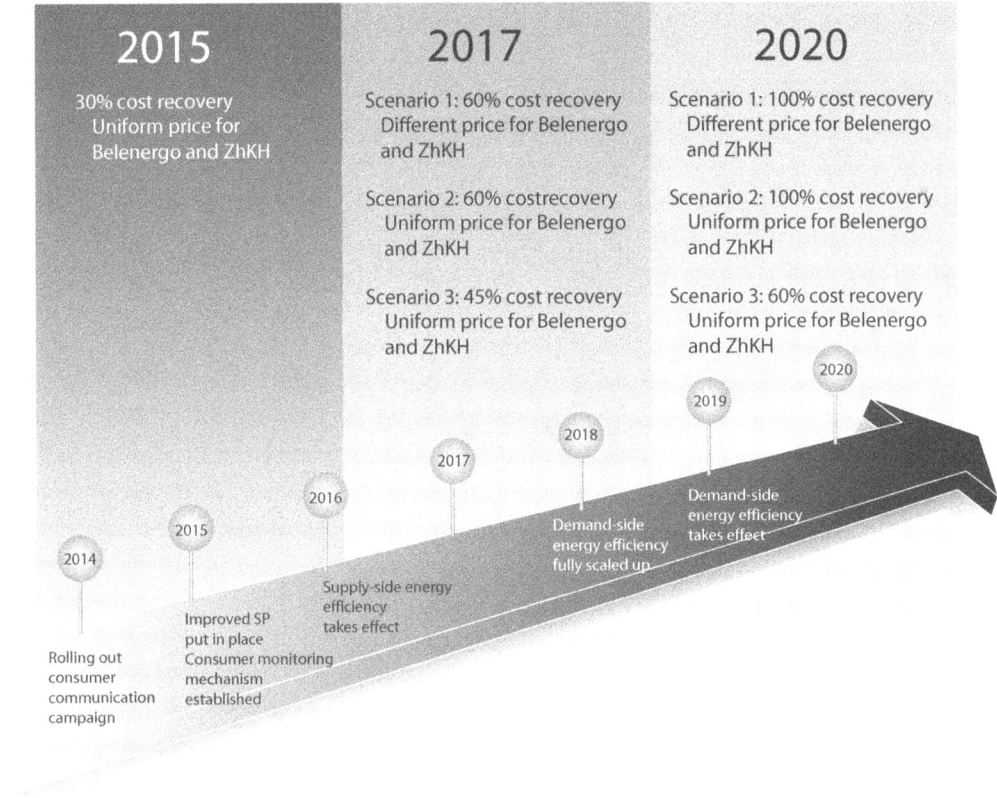

2015

30% cost recovery
Uniform price for
Belenergo and ZhKH

2017

Scenario 1: 60% cost recovery
Different price for Belenergo
and ZhKH

Scenario 2: 60% costrecovery
Uniform price for Belenergo
and ZhKH

Scenario 3: 45% cost recovery
Uniform price for Belenergo
and ZhKH

2020

Scenario 1: 100% cost recovery
Different price for Belenergo
and ZhKH

Scenario 2: 100% cost recovery
Uniform price for Belenergo
and ZhKH

Scenario 3: 60% cost recovery
Uniform price for Belenergo
and ZhKH

2020

2019

2018

2017

2016

2015

2014

Demand-side
energy efficiency
takes effect

Demand-side
energy efficiency
fully scaled up

Supply-side energy
efficiency
takes effect

Improved SP
put in place
Consumer monitoring
mechanism
established

Rolling out
consumer
communication
campaign

Belarus Heat Tariff Reform and Social Impact Mitigation
http://dx.doi.org/10.1596/978-1-4648-0696-4

Overview of the District Heating Sector in Belarus

Introduction

This section contains a brief overview of the supply and demand characteristics of the district heating sector in Belarus, the characteristics of the main service providers, the legal and regulatory framework, and the levels of tariffs and subsidies in the sector.

Demand and Supply Characteristics

Sixty one percent of the households in Belarus—81 percent of urban and 14 percent of rural households—rely on district heating for heat supply. Consumption across all sectors totaled roughly 47 million GCal in 2012 and has been relatively flat in recent years. Residential customer consume 34 percent of the heat produced. Industrial and other customers (for example, health care and agriculture) consume 46 percent and 20 percent, respectively.

Approximately half of the heat is produced in combined heat and power plants (CHPs) and half in boiler houses. Natural gas is the primary fuel, with very small amounts of biofuels and waste and fuel oil also used. The share of natural gas use as fuel increases with growth of boiler house capacity. Small boiler houses generally use solid fuels, medium-capacity boiler houses— natural gas and solid fuels; natural gas prevails in large boiler houses.

Gas is imported mainly from the Russian Federation: 21.6 billion cubic meters were imported in 2010 (about 63 percent of primary energy consumption[1]), of which more than one-third was consumed in district heating. Figure A.1 illustrates heat production by source for 2007–12.

This appendix is based on a background study by Denzel Hankinson and a political economy analysis by Izabela Leao and Ecaterina Canter.

Figure A.1   Heat Production, by Source, 2007–11

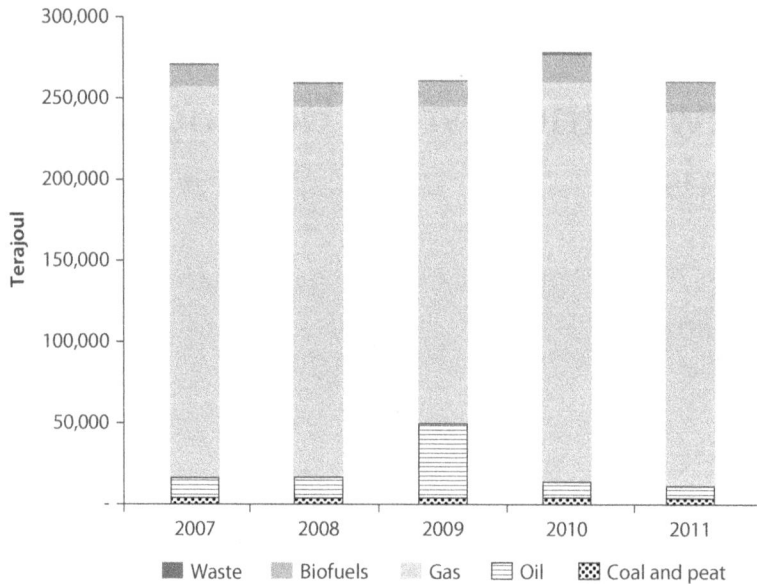

Source: Energy Charter Secretariat 2013.

Service Providers

There are two main providers of DH in Belarus—Belenergo and municipally owned communal service companies called ZhKHs. A small portion (less than 1 percent) of heat demand is met by small private DH companies.

Belenergo

Belenergo is a vertically integrated, state-owned company that supplies heat and electricity through six regional companies. It supplies the vast majority of all electricity consumption and roughly 50 percent of district heating end-use consumption in Belarus. Figure A.2 illustrates the structure of Belenergo.

As of 2013, the installed capacity of all power and heating facilities of Belenergo SPA was about 8,220 megawatts, which included 3,988 megawatts of condensing power plants, 3,982 megawatts of large CHPs, 182 megawatts of small CHPs (under 50 megawatts), 12 megawatts of local fuel-fired CHPs, and 33 megawatts of small hydropower plants.

ZhKHs

ZhKHs are the other major supplier of district heating services. Roughly 120 ZhKHs provide a number of services to tenants, including district heating, running water, and sewage services. ZhKHs provide district heating services in urban areas, small towns, and rural areas where Belenergo does not operate. ZhKHs do not supply electricity.

Figure A.2 Organization of Belenergo

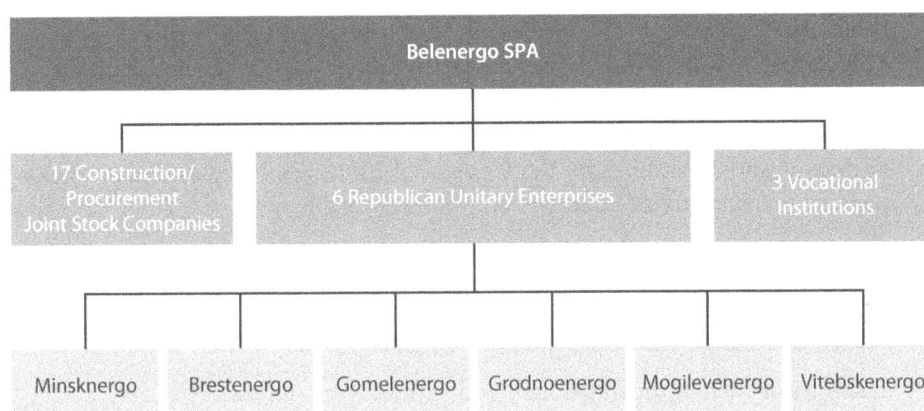

Source: energo.by, Energy Charter Secretariat.
Note: SPA = State Production Association.

Table A.1 Capacity of Boiler Houses Operated by ZhKHs

Gcal/hour

	Total	Less than 0.5	0.5–3	3–10	10–100	100 and above
Availability as of the reporting year end	3,691	1,000	1,864	587	237	3
Aggregate capacity of the boiler houses	12,156.7	357.4	2,798.4	2,971.3	5,694.2	335.3
Number of boiler houses	3,691	1,000	1,864	587	237	3
Fuel type						
Solid	2,781	874	1,551	313	43	0
Liquid	12	1	8	2	1	0
Including mazut	2	0	0	1	1	0
Fuel gas	897	125	305	272	192	3
Number of installed boilers, including CHPs	10,484	3,729	5,104	1,229	417	5
Fuel						
Solid	6,450	3,205	2,822	356	67	0
Mazut	161	5	85	50	21	0
Fuel oil	33	5	18	10	0	0
Fuel gas	3,868	611	2,134	793	325	5

Source: Ministry of Housing and Utilities.
Note: CHPs = combined heat and power plants.

As of 2014, the ZhKHs collectively had 3,691 boiler houses with total installed capacity of 12,157 Gcal/h, as well as 15,800 kilometers of heat networks. Small and medium boiler—with capacities under 3 megawatts—generated 78 percent of the heat but represent only 26 percent of total installed capacity. Table A.1 summarizes the characteristics of boilers houses operated by the ZhKHs.

Belarus Heat Tariff Reform and Social Impact Mitigation
http://dx.doi.org/10.1596/978-1-4648-0696-4

Policy and Regulation

Policy and regulation in the Belarus' district heating sector is highly centralized. The central government, which establishes the sector's policy and legal framework, sets the rules, and determines how resources are distributed. Service provision is regulated primarily through decrees and directives issued by the President of Belarus and resolutions by the Council of Ministers, based on recommendations by the Ministry of Economy and Ministry of Energy.

Main Entities Responsible

Figure A.3 summarizes the roles of government of Belarus (GoB) individuals and entities in setting policy and regulating the DH sector.

The roles of each of the individuals and entities are as follows:

- The president of Belarus has final approval of changes in residential tariffs proposed by the Council of Ministers. The president approves or revises the tariff proposals submitted by the Council of Ministers and puts the changes into force through presidential decrees.
- The Council of Ministers approves or revises proposed changes in tariffs put forward by the Ministry of Economy. Unlike tariffs for nonresidential

Figure A.3 Responsibilities for Policy and Regulation in the District Heating Sector

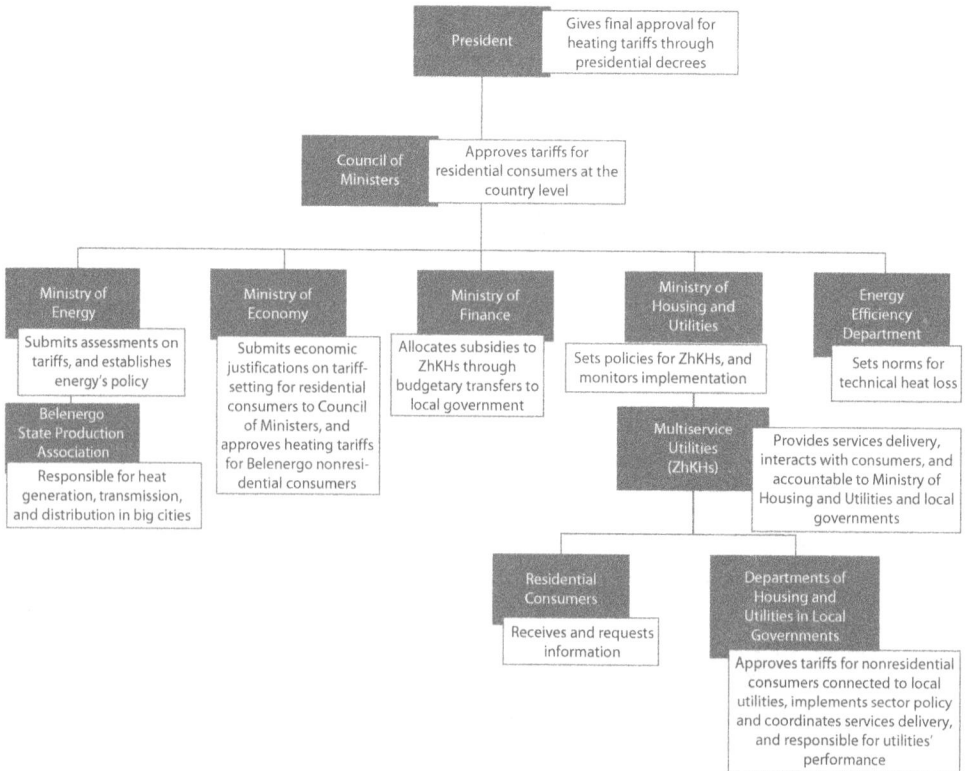

consumers, which vary across the country and are approved at different levels by the Ministry of Economy and local governments, the residential heating tariffs are the same throughout the country and approved at the Council of Ministers level. To review and discuss the tariffs increases in the Council of Ministers' sessions, all sectoral ministers and relevant sector enterprises must express their positions and agree on the tariff increase before the official deliberation.

- The Ministry of Economy is principally responsible for the economic analysis needed to determine tariffs for residential consumers and for Belenergo's non-residential consumers. The Ministry of Economy conducts economic assessments on the need for tariffs increase to residential customers, and submits them to the Council of Ministers. The Ministry of Economy also establishes tariffs for nonresidential consumers connected to Belenergo's heat supply networks, based on information provided by the Ministry of Energy.

- The Ministry of Energy is responsible for policy and coordination of Belenergo's organizations, and holds them accountable for their performance. It provides economic estimates to the Ministry of Economy as the basis for setting tariffs for Belenergo's nonresidential customers, and performs the analysis necessary for setting tariffs for nonresidential consumers connected to Belenergo's networks.

- The Ministry of Finance allocates funds for investment in, and subsidies to, the heating sector. The Ministry of Finance is responsible for financial transfers, through the Ministry of Housing and Utilities, to Oblast governments, which then distribute these to local multiservice utility companies based on their subsidy needs.

- The Ministry of Housing and Utilities sets policies for the ZhKHs and monitors implementation of those policies.

- Local governments are responsible in sector coordination and service provision. Oblast governments are closely involved in sector coordination, while municipal/rayon governments are responsible for service provision through the ZhKHs. The local government's role in the heating tariff reform is to support and monitor the ZhKHs. At the Oblast level, the Department of Housing and Utilities in the Oblast Executive Committee coordinates the sector activities in the Oblast. At the rayon and city levels, the Department of Housing and Utilities of the Executive Committee are responsible for the operations of the ZhKHs based on Oblast governments' guidance. The rayon- or city-level governments do not have authority to establish regulations or benchmarking in their jurisdiction. The local ZhKHs file regular reports to rayon or city governments on their performance. In addition, Oblast Executive Committees establish heating tariffs for Oblasts' nonresidential consumers.[2]

- The Energy Efficiency Department indirectly participates in the tariff setting by approving norms for technical heat losses. The Energy Efficiency Department is part of the Standardization Committee, which is directly accountable

Table A.2   National Cost-Recovery-Level Targets for Heat and Electricity Sectors, 2012–15

Energy type and prime cost	2012 (actual)	2012 (actual)	2013	2014	2015
Heat (distributed by Belenergo suppliers, %)	21.4	17.2	18.7	21.0	30.0
Prime cost of 1 Gcal of heat, BYR/Gcal	202,185.5	329,273.9	359,649.6	406,217.8	453,138.4

Source: Program for Energy Sector Development until 2016.

to the Council of Ministers. The department gives estimates on energy efficiency targets to the Ministry of Housing and Utilities, which splits them between oblasts, sectors, and enterprises. The enterprises are responsible for the implementation of a range of energy efficiency measures, including fuel savings. The Energy Efficiency Department supervises the implementation results at large enterprises, while Oblast Executive Committees are responsible for supervision of smaller enterprises.

Principal Laws and Regulations

The GoB's Strategy for Energy Potential Development (Resolution 1180) sets national targets for the energy sector up until 2020. The overall objective of the strategy is to ensure Belarus' energy independence and promote the efficient use of energy resources. The GoB targets relevant to the DH sector include the following:

- Increasing the share of domestic fuel in the energy mix to 28–30 percent by 2015 and 32–34 percent by 2020, reducing dependence on imported natural gas
- Reducing the share of natural gas in the energy balance to 64 percent in 2015 and to 55 percent by 2020
- Reducing the energy intensity of GDP by 50 percent by 2015, and by 60 percent by 2020 (from 2005 levels)
- Phasing out subsidies and cross subsidies
- Restructuring heat tariffs.

In line with the strategy, the GoB has enacted a number of DH sector-specific policies and legislation described below.

State Program for the Development of the Belarussian Energy System until 2016 (Resolution 194)

To achieve the goals set forth in Resolution 1180, the GoB introduced Resolution 194, a program which aims to increase the efficiency and reliability of the Belarusian energy system. The program lays out specific cost-recovery targets for Belenergo's DH service (appendix table A.3) and other policy objectives related to the DH sector such as developing and modernizing DH networks, tariff reforms, not limited to phasing out cross subsidies, and improvements to the managerial, regulatory, and legal framework of the DH sector. Specific strategies highlighted in the program include the following:

- Constructing networks with high-insulation and automated control systems
- Annual replacement of 100–120 kilometers of pipelines for Belenergo and 550–660 kilometers of pipelines for ZhKHs
- Increasing local fuels in the heat fuel mix
- Decreasing network losses
- Transferring heat loads from ZhKH-owned networks to Belenergo-controlled networks
- Increasing the transparency of tariff setting for all customer classes
- Introducing in-floor radiant heating in new building stock.

Program for Housing the Utilities of the Republic of Belarus 2015 (Resolution No. 97)

In 2013, the GoB passed Resolution 97 to increase the efficiency and reliability of services provided by the ZhKHs. The program included strategies and policy goals for DH services of the GoB at the municipal level such as:

- Replacing 3.8 thousand kilometers of the current DH network, using insulated pipes
- Replacing not less than 9,000 heating pumps
- Replacing old boilers
- Decentralizing DH systems and instead introducing more localized DH supply
- Increasing the share of local fuels in the DH mix from 34.9 percent in 2012 to 54.4 percent in 2015
- Increasing utility tariffs, reduction of cross subsidies, and tariff reform that reflects the area of the premises
- Reducing losses of the heat network by 6.7 percent by 2016 from 2010 levels (19 percent heat loss).

Table A.3 describes the GoB's targets for replacing old and poorly performing pipes in the DH network.

Table A.3 Targets for Replacement of District Heating Pipes, by Region
kilometers

Location	2011 (actual)	2012 (actual)	2013	2014	2015	Total
Brest region	125.1	125.1	124.0	115.0	100.0	589.2
Vitebsk region	81.7	110.1	120.0	110.0	106.0	527.8
Gomel region	116.1	120.3	125.0	125.0	130.0	616.4
Grodno region	97.9	98.8	100.0	100.0	100.0	496.7
Minsk region	134.5	143.3	140.0	147.0	150.0	714.8
Mogilev region	132.7	126.7	130.0	130.0	145.0	664.4
Minsk city	50.0	52.0	30.0	45.0	45.0	222.0
Total	738.0	776.3	769.0	772.0	776.0	3,831.3

Source: Resolution 97, Council of Ministers.

Laws Related to the District Heating Sector

The GoB has passed a number of laws on tariff reform in accordance with its policy objective of decreasing subsidies and cross-subsidies for the residential DH sector. Key legislation includes the following:

- *The Law on Pricing No.255-3:* This sets out the framework for price regulation and the state bodies that are responsible for regulation and governance of sectors of national interest, including the DH sector.
- *Presidential Decree No. 72:* This defines regulatory responsibilities of state bodies of the republic of Belarus for pricing of goods and services, including tariff-setting responsibilities for district heating services.
- *Presidential Decree No. 550:* This establishes procedures for financial management of public utilities, including the processes for setting residential tariffs, fiscal planning and accounting associated with technical maintenance and new capital expenditures and financing of capital investments. In line with the government's plans to gradually eliminate cross-subsidies and increase cost-recovery levels of residential DH service, the decree mandates an annual increase of residential tariffs (including DH) of up to 5USD per year; any additional increase of tariffs must be approved by the president. Furthermore, residential tariffs are indexed quarterly by the growth of household income that does not exceed the nominal growth rate of wages.
- *Resolution of the Council of Ministers No. 1166:* This sets DH tariffs for 2014, for the DH sector for subsidized and nonsubsidized customers.

Tariff and Subsidies in the District Heating Sector

Residential tariffs are well below the cost of service in Belarus. Since 2005, the cost of producing heat has doubled in real terms, while residential DH tariff increases have remained relatively flat as nominal increases by the GoB have been largely eroded by inflation. The cost-recovery rate for residential DH services provided by the ZhKHs declined from 45.3 percent in 2005 to 14.5 percent in 2011 and from 74.8 percent to 21.4 percent for Belenergo. The difference in cost-recovery rates are associated with Belenergo's economies of scale and use of highly efficient CHPs and the ZhKH's inefficient and outdated boilers. Appendix Figure A.4 shows the trend cost of production, cost-recovery levels, and residential DH tariffs since 2005.

The Republic of Belarus relies heavily on natural gas imports from Russia to meet domestic energy demand. While the country still imports natural gas at below European market prices, the cost of a terra cubic meter of natural gas has risen by about 1,300 percent since 2006, from 100 BYR to 1,400 BYR. This sharp increase has been offset only slightly by the reduction in technical losses in the transmission and distribution systems—currently 15 percent for ZhKH and 10 percent for Belenergo. Since natural gas imports account for 60 percent of DH production, the GoB has increased direct subsidies and cross-subsidies to

Figure A.4 Cost of Production, Cost-Recovery Levels, and Residential Tariffs of the District Heating Sector, 2005–12

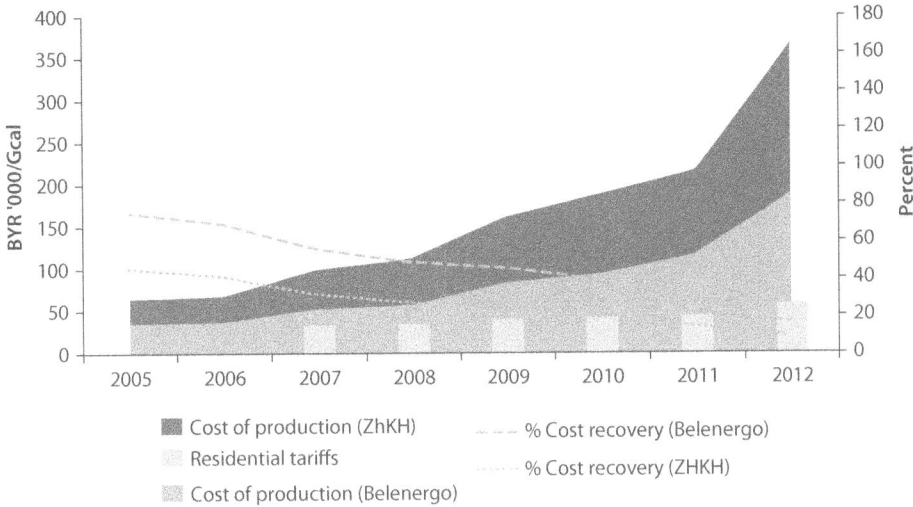

Legend:
- Cost of production (ZhKH)
- Residential tariffs
- Cost of production (Belenergo)
- % Cost recovery (Belenergo)
- % Cost recovery (ZHKH)

Source: Ministry of Economy, Ministry of Housing and Utilities.

Figure A.5 Rising Costs of Natural Gas Imports, 2005–12

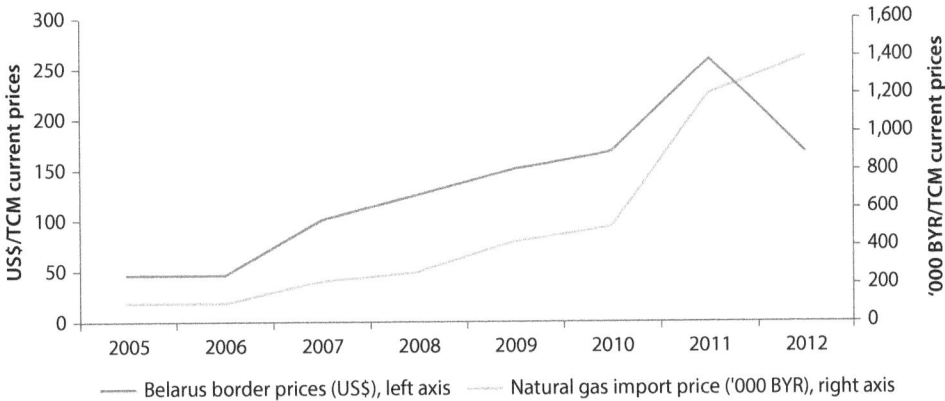

Legend:
- Belarus border prices (US$), left axis
- Natural gas import price ('000 BYR), right axis

Source: Ministry of Economy, Organization for Economic Cooperation and Development.

compensate for declining cost-recovery levels of DH providers. Figure A.5 shows the rapid increase of natural gas imports since 2005, and figure A.6 shows how direct subsidies have increased from 0.5 percent of gross domestic product (GDP) in 2005 to about 2 percent in 2012.

Direct subsidies are insufficient to compensate for the sharp increase in production costs, so cross-subsidies between nonresidential DH customers and electricity are used to cover DH providers' revenue requirements. Belenergo, which currently achieves about 20 percent cost recovery from residential heat consumers, does not receive state subsidies and makes up the entire shortfall by cross-subsidization.[3] As a result, Belenergo's electricity and nonresidential heat consumers pay tariffs that are substantially higher than cost, in order to subsidize residential heat consumers. This is reflected in the growing gap between tariffs paid by residential and nonresidential consumers illustrated in figure A.7.

On the other hand, the ZhKHs have compensated for the falling value of residential revenue with substantial increases in state subsidies, together with cross-subsidization from nonresidential consumers, as shown in figure A.8.

Figure A.6   Fiscal Cost of District Heating Subsidies, 2005–12

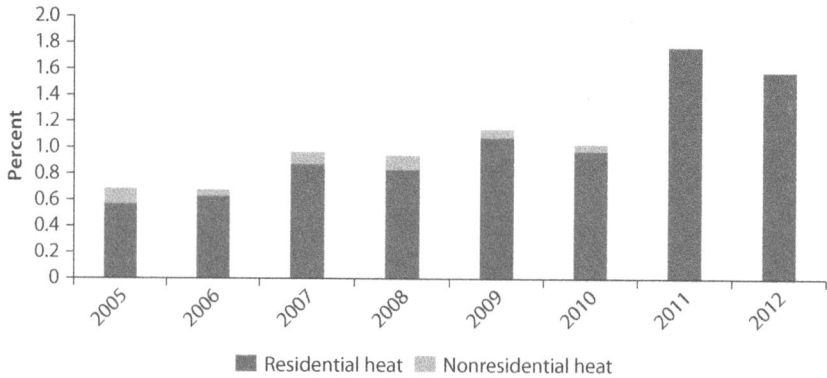

Source: Ministry of Finance, Belenergo, ZhKH, Beltopgas, and World Bank estimates.

Figure A.7   Amount of Cross-Subsidies for Residential District Heating (Belenergo), 2007–12

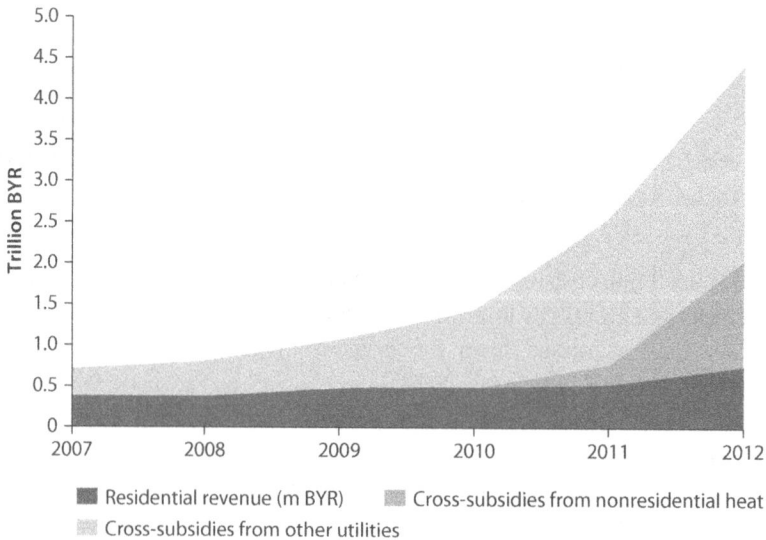

Source: Ministry of Economy.

Figure A.8 Amount of Cross-Subsidies for Residential District Heating (ZhKH), 2005–13

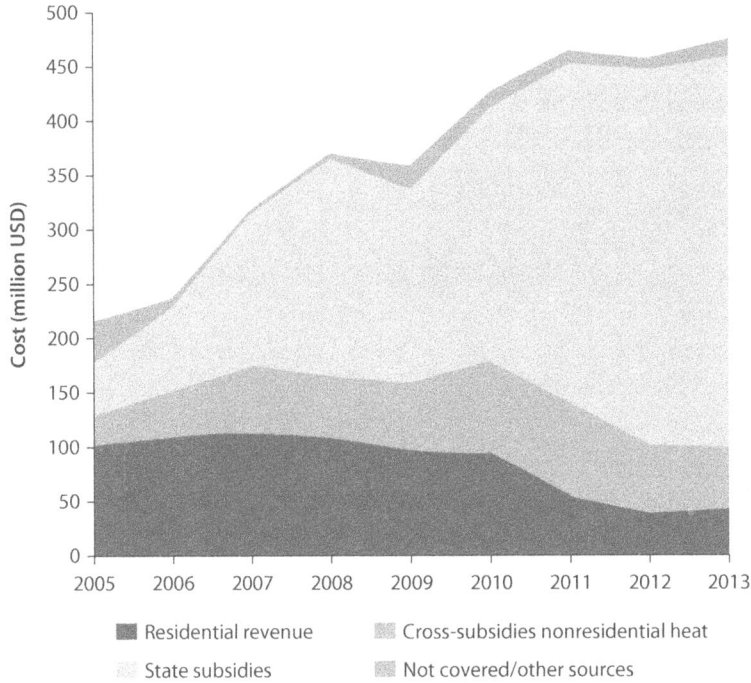

Legend:
- Residential revenue
- Cross-subsidies nonresidential heat
- State subsidies
- Not covered/other sources

Source: Ministry of Housing and Utilities.

Notes

1. In 2010, 20.7 percent of primary energy consumption was met from domestic sources: oil, biofuel and waste, peat, and natural gas.

2. Under the recent Presidential Decree N550, the local authorities will be responsible for tariff setting for certain utility services, such as water provision, buildings' maintenance, and waste management, which were formerly the national government's prerogative.

3. Resolution 220 mandates cross-subsidization between types of utility service and between customer classes to achieve cost recovery.

The Distributional Impact of Tariff Reform on Households and Industries

Introduction

Using Household Budget Survey data from 2007 to 2012, the annex provides a descriptive analysis of the energy expenditure patterns of Belarusian households. In addition, a simulation is conducted to estimate the distribution impact of increasing district heating tariffs in the residential sector. Finally, the annex examines how reducing the cross-subsidization from nonresidential electricity to residential heat could improve the competitiveness of industries in Belarus. The analysis benchmarks the unit energy cost of per dollar manufacturing value added of Belarus against 32 countries in the region.

Household Energy Expenditure Patterns

Analysis of the HBS 2012 indicates that an average of 32 percent of communal service expenditures is spent on district heating. When broken down by quarter, district heating expenditures represents 45 percent, 18 percent, 15 percent, and 40 percent of the total communal service expenditures for the 1st, 2nd, 3rd, and 4th quarter, suggesting a strong correlation between colder temperatures in Q1 and Q4 and district heating consumption

Energy expenditures are defined as the sum of expenditures on district heating, electricity, gas, firewood, turf (or peat), coal, and others. The share of household income on energy expenditure is inversely related to household income levels. In 2012, the bottom income quintile (that is, the poorest) household spent 4.9 percent of household income on energy expenditure, while the 5th income quintile (that is, the richest) spent 1.6 percent. Rural households spend a higher proportion of their income on energy expenditures than urban households. In 2012, the rural bottom 20 percent household spent 4.4 percent of household income on energy expenditures, while the urban bottom quintile spent only 3.8 percent. The overall aggregated trend for both rural and urban households is shown in figure B.1.

This appendix is based on a background study by Fan Zhang, Bonsuk Koo, and Karuna Phillips.

Figure B.1 Annual Share of Household Income on Energy Expenditure, by Income Quintile, 2007–12

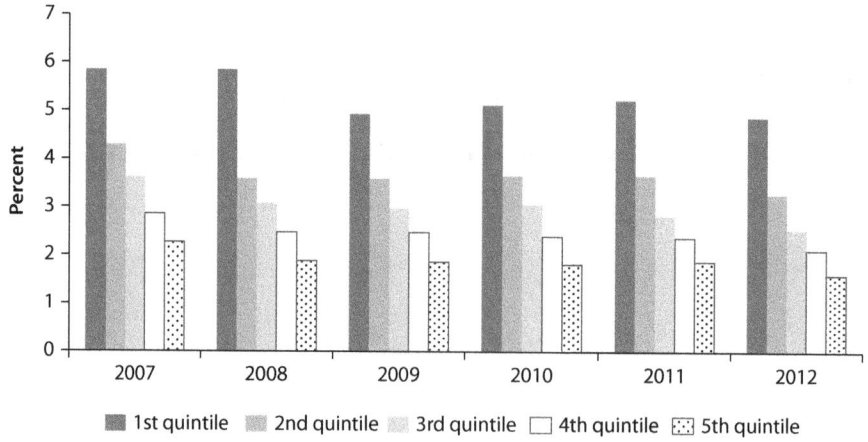

Source: Belarus Household Budget Survey from 2007 to 2012.
Note: Income and expenditure in 2007 value is adjusted by Consumer Price Index.

Figure B.2 Share of Energy Expenditure, by Fuel Source, 2007–12

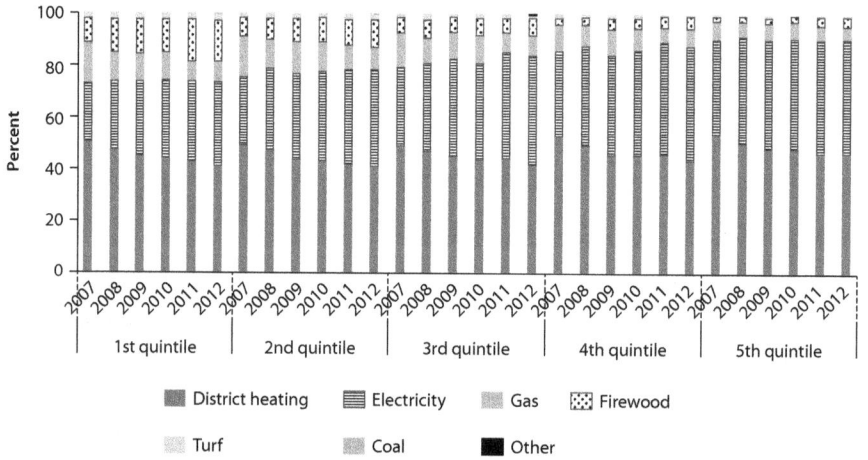

Electricity and district heating account for the largest share of energy expenditure. Figure B.2 indicates that higher income households spent more on electricity and district heating as a proportion of total energy expenditure than low-income households. Lower-income households spent more on alternative fuels such as firewood, turf, and coal. The share of energy expenditures for district heat has decreased over the years across all income quintiles.

Expenditure on district heating has decreased across all income quintiles since 2007. The decreasing trend in expenditure on district heating is in line with heating tariff decrease over time as shown in figure B.3. Rural households have decreased district heating expenditures more than urban households. The

Figure B.3 Annual Average Residential Heat Tariff, 2005–07

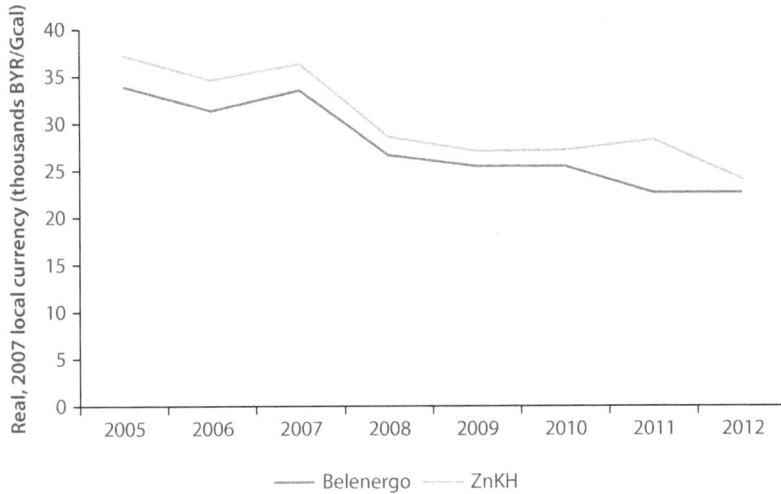

Source: Ministry of Economy.
Note: Income and expenditure in 2007 price is adjusted by consumer price index.

bottom quintile, 2nd quintile, 3rd quintile, and 4th quintile in rural areas have decreased district heating expenditures by 4.5 percent, 4.3 percent, 4.3 percent, and 5.4 percent, respectively. The bottom, 2nd, 3rd, and 4th quintile in urban areas have decreased district heating expenditures by only 2.9 percent, 2.4 percent, 2.9 percent, and 3.4 percent, respectively.

Electricity expenditures have constantly increased over time and across all income quintiles. Even though electricity expenditures have been increasing for the last seven years, the share of household income on electricity has been constant due to the fact that household incomes have been increasing as well.

In contrast to district heating and electricity, gas and alternative fuels expenditures only account for a small part of household energy expenditures. Household gas expenditures have decreased as has the proportional share of household income spent on gas. Alternative fuel expenditures have increased, though their relative share of household income is rather small. However, the bottom 40 percent households have increased their expenditure on alternative fuels quite rapidly. The bottom income quintile spent 4,121 BYR on alternative fuel expenditures in 2007, but it rose by over 60 percent to 6,645 BYR in 2012.

Urban and poor households allocate more of their budget on district heating expenditure than rural and rich households. Figure B.4 demonstrates the district heating expenditure budget share difference between urban and rural has widened over time, even though the share of household income on energy expenditures in 2012 is lower than in 2007. In 2012, more than 50 percent of energy expenditure was spent on district heating across all income quintiles in urban area, while the share of energy expenditure on district heating is slightly above 35 percent among rural households. The gap in the share of income on spent district heating between rural and urban areas has widened over time.

Belarus Heat Tariff Reform and Social Impact Mitigation
http://dx.doi.org/10.1596/978-1-4648-0696-4

Figure B.4 Share of Energy Expenditure, by Fuel Sources, 2007–12

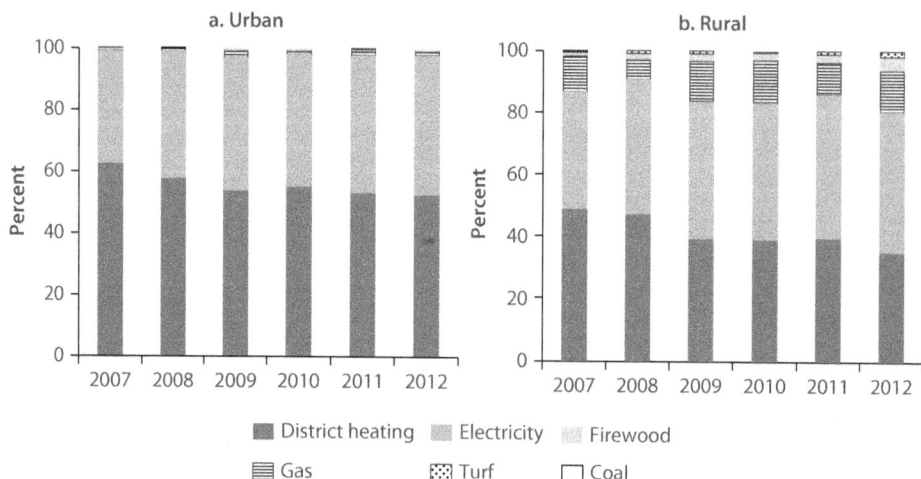

Source: Calculation based on the HBS from 2007 to 2012.
Note: Income and expenditure in 2007 value is adjusted by consumer price index.

Distributional Impact of District Heating Tariff Increase

Assuming zero heat price elasticity, three tariff increase scenarios are explored to estimate the distribution impact of tariff reform. Table B.1 summarizes the scenarios, which were devised based on different district heating pricing regimes and cost-recovery goals. A uniform pricing regime is to provide district heating at the same price. Under a differentiated price regime, Belenergo and ZnKHs provide district heating at the different price reflecting their production cost.

The results of simulation indicate that rural poor households are most vulnerable under the differentiated price scheme. In contrast, the urban poor are most vulnerable under the uniform pricing regime (see figure B.5).

Rural poor households are most vulnerable under the differentiated price regime. Figure B.5 indicates that the average share of household income on district heating would be 17.1 percent for the rural bottom income quintile and 14.7 percent for the urban bottom income quintile at 100 percent of cost-recovery levels. Due to economies of scale, ZhKH, the main district heat provider for households in small cities and rural areas, will need to set higher district heat tariffs than Belenergo, the main provider for households in Minsk and large cities. Thus, under the differentiated price regime, rural households would have to bear more a financial burden resulting from tariff increases under a differentiated pricing regime than urban households.

Urban poor households are most vulnerable under the uniform pricing regime. As shown in figure B.6, urban and rural poor households would spend 19.4 percent and 14.6 percent of household income, respectively, on district heating at full cost-recovery levels.

Belarus Heat Tariff Reform and Social Impact Mitigation
http://dx.doi.org/10.1596/978-1-4648-0696-4

Table B.1 Tariff Reform Scenarios, 2015–20

	2015		2017		2020	
	Cost recovery goal (%)	Pricing	Cost recovery goal (%)	Pricing	Cost recovery goal (%)	Pricing
Scenario 1	30	Uniform	60	Differentiated	100	Differentiated
Scenario 2	30	Uniform	60	Uniform	100	Uniform
Scenario 3	30	Uniform	45	Uniform	60	Uniform

Figure B.5 Share of Household Income on District Heating in First Scenario

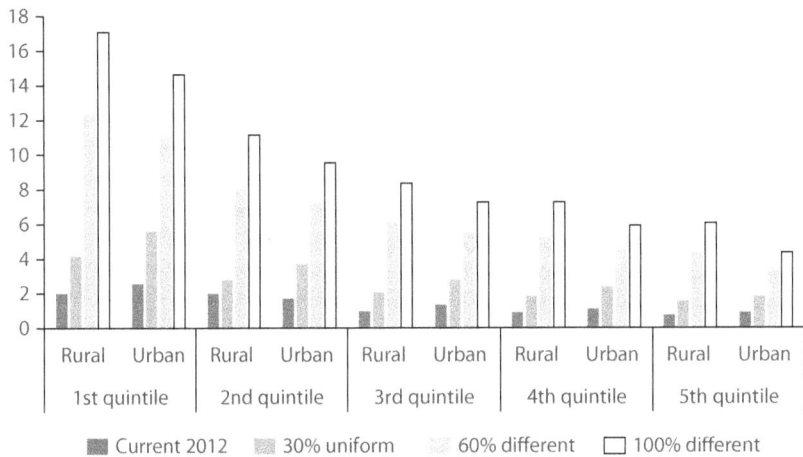

Source: Calculation based on 2012 HBS.

Figure B.6 Share of Household Income on District Heating in Second Scenario

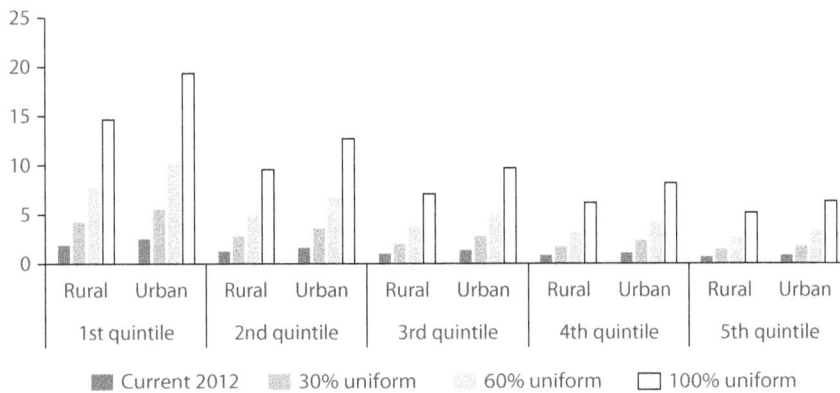

Source: Calculation based on 2012 HBS.

The share of income on district heating spikes during the heating season in the 1st and the 4th quarter. The seasonal fluctuation of district heating budget share is shown in figure B.7.

Belarus Heat Tariff Reform and Social Impact Mitigation
http://dx.doi.org/10.1596/978-1-4648-0696-4

Figure B.7 Share of Average Quarterly Income on District Heating of the Bottom 20 Percent Household

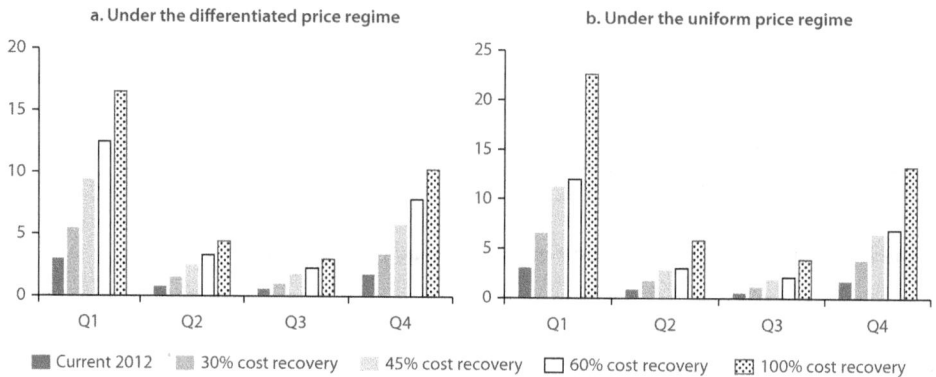

a. Under the differentiated price regime

b. Under the uniform price regime

■ Current 2012   ▨ 30% cost recovery   ▨ 45% cost recovery   ☐ 60% cost recovery   ▦ 100% cost recovery

Source: Simulation based on 2012 HBS.

Cross-Subsidization and Industrial Competitiveness

Underpriced residential heat has been financed largely by the cross-subsidies from Belnergo's nonresidential electricity sales. The direct budgetary subsidies from local government to ZhKHs are less than 33 percent of fiscal cost and the rest of fiscal cost is financed by nonresidential sector. To measure the impact of cross-subsidization removal on industry, the energy cost of one US dollar value added for each industry will be benchmarked taking into account industrial energy tariffs, industrial energy intensity, and the ratio of electricity and gas consumption of the production processes. Energy consumption data are obtained from the IEA energy balance and statistics database; the industrial value added data are obtained from UNIDO INDSTAT 4. Energy tariffs are obtained from ERRA tariff database and Eurostat.

Reducing nonresidential electricity price to cost-recovery level could improve the industrial competitiveness of Belarus. Total industrial energy cost per US dollar value added in Belarus at current gas and electricity price is ranked 17th among 32 European countries, as shown in figure B.8. If nonresidential electricity price decreases to cost-recovery level, Belarus would be ranked 13th holding gas price at current level.

Four industries would gain the most from the removal of cross-subsidies. These are wood production, paper and pulp, textile, and food and tobacco. The energy intensity of the wood and wood production industry in Belarus is 26th among 32 countries, while the energy cost per one US dollar of value added in the wood and wood production industry is ranked 24th, as shown in figure B.9. The average energy cost of Poland, Romania, and Lithuania in the analysis is 38, 34, and 31 cents, respectively. Thus, the energy cost per one dollar value added in the Belarusian wood production industry is higher than these three countries

Figure B.8 Unit Energy Cost of Manufacturing, 2009

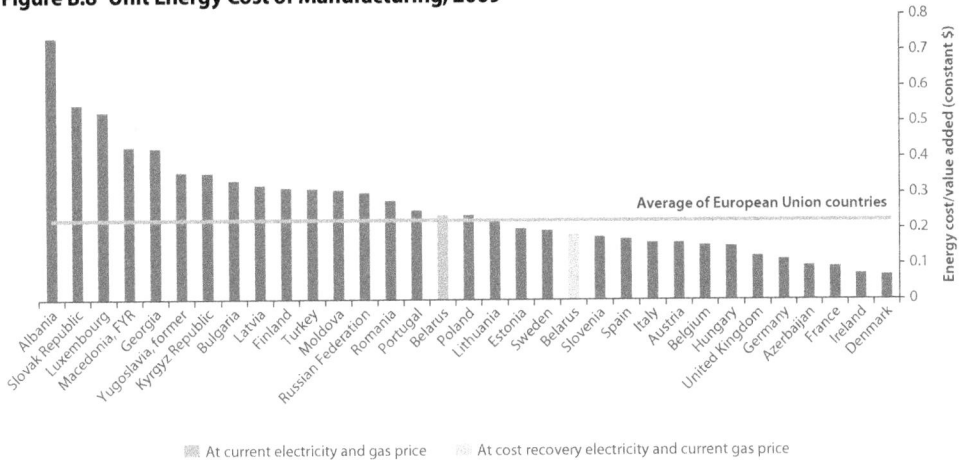

Source: Calculation based on IEA Energy Balance and Statistics, UNIDO INDSTAT 4, ERRA Tariff and Eurostat Databases.

Figure B.9 Unit Energy Cost of Wood Production, 2009

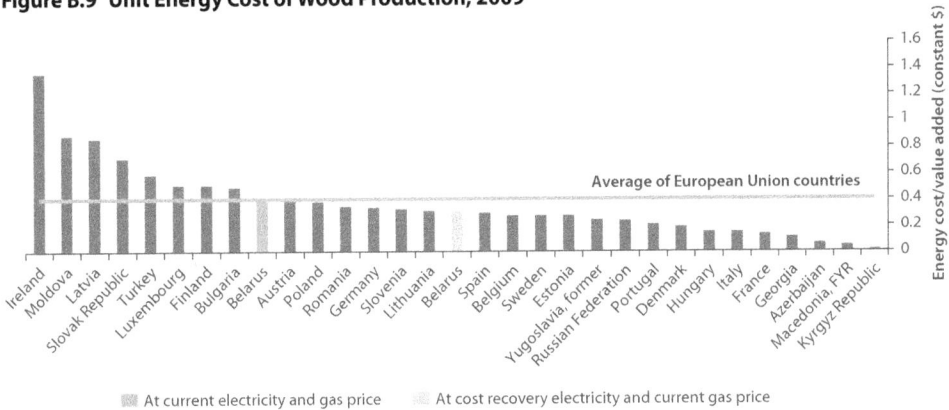

Source: Calculation based on IEA Energy Balance and Statistics, UNIDO INDSTAT 4, ERRA Tariff and Eurostat Databases.

Figure B.10 Unit Energy Cost of Paper, Pulp, and Print, 2009

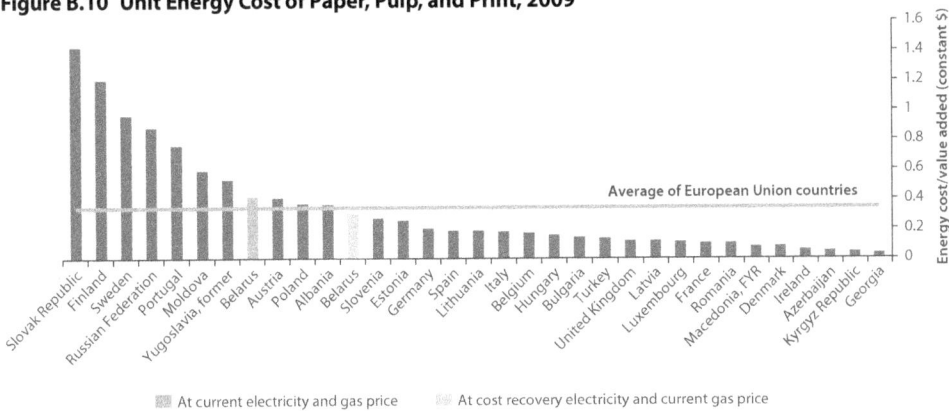

Source: Calculation based on IEA Energy Balance and Statistics, UNIDO INDSTAT 4, ERRA Tariff and Eurostat Databases.

Belarus Heat Tariff Reform and Social Impact Mitigation
http://dx.doi.org/10.1596/978-1-4648-0696-4

Figure B.11 Unit Energy Cost of Textile and Leather Industry, 2009

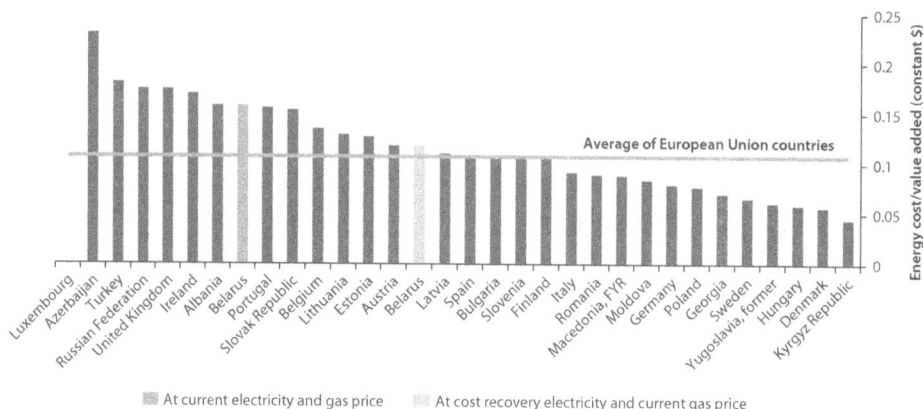

Source: Calculation based on IEA Energy Balance and Statistics, UNIDO INDSTAT 4, ERRA Tariff and Eurostat Databases.

Figure B.12 Unit Energy Cost of Food and Tobacco Industry, 2009

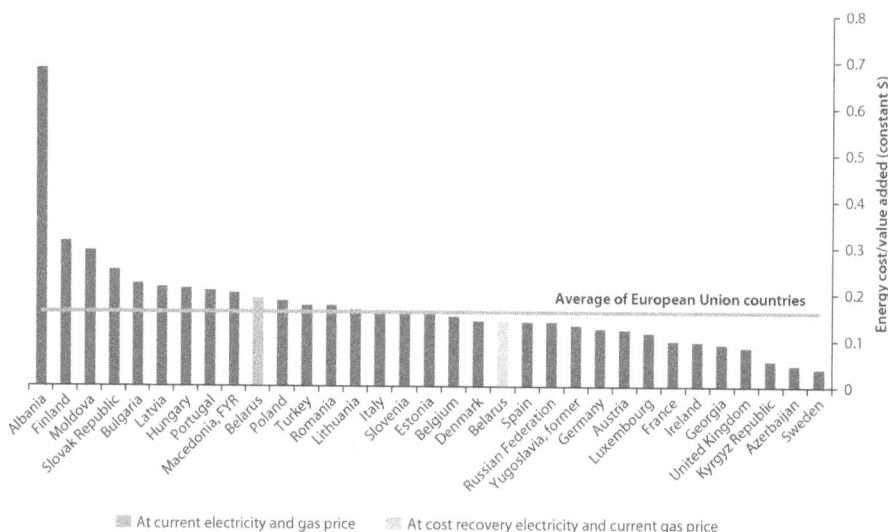

Source: Calculation based on IEA Energy Balance and Statistics, UNIDO INDSTAT 4, ERRA Tariff and Eurostat Databases.

above. When electricity price is reduced to cost-recovery level, then the energy cost of Belarus could be lower than those countries.

Energy intensity and energy costs for the paper, pulp, and print industry at current gas and electricity price is 25th among 32 countries, as shown in figure B.10. When electricity price is adjusted to cost-recovery level, the energy cost of the paper and pulp industry would be reduced by approximately 28 percent.

Energy costs for the textile and leather industry in Belarus are also high. Per unit energy cost and the energy intensity is ranked 26th and 29th at current electricity and gas price, respectively, as shown in figure B.11. The unit energy cost of the textile and leather industry in Belarus is higher than the average of 32 countries by approximately 30 percent. If electricity price is adjusted to cost-recovery level holding gas price constant, the energy cost of the textile and leather industry in Belarus would be decreased to 12 cents per dollar value added and will be ranked 19th among the 32 countries.

Food and tobacco industry in Belarus spent 19 cents to generate one dollar of value added in 2009. When nonresidential electricity prices are decreased to cost-recovery level, the energy cost for the food and tobacco industry in Belarus would be reduced by 26 percent. Belarus was ranked 23rd among 32 countries at current electricity and gas price, as shown in figure B.12, and would be ranked 14th among 32 countries at cost-recovery electricity and current gas price.

Methodologies of Focus Group Discussions and In-Depth Interviews

Introduction

Qualitative social impact analysis of district heating tariff increase was conducted by focus group discussions (FGD), in-depth interviews with experts (IDI), and ethnographic interviews.

FGDs with households were conducted to provide information on population's energy consumption patterns and assessment of the district heating in Belarus, household's affordability of heating tariff increase and their coping strategies in adjusting to increasing heating expenditures, existing social assistance mechanisms, and energy efficiency and renewable energy strategies of population, households' awareness about current quality of services, tariff-setting process, and so forth.

IDIs with experts were conducted to capture experts' assessment of the situation in the district heating sector, assessment of main tendencies of district heating, existing and prospective energy efficiency measures, affordability of existing and increasing heating expenditures for poor and vulnerable people, and existing and prospective social measures to mitigate negative impacts of heating tariffs increase.

Focus Group Discussions

Eighteen FGDs were conducted for the purpose of the research: 2 pilot FGDs and 16 FGDs for main field stage. The composition of the focus group samples are summarized in table B.1. The structure of FGDs was organized according to the following criteria:

- Income: 12 FGDs included households in the bottom two income quintiles (no more than 2 million BYR per capita)—the most vulnerable groups. Six FGDs were also conducted with middle-income households (the third and fourth income quintile—from 2 million to 3.5 million BYR per capita).

This appendix was contributed by SATIO.

Table C.1 Focus Group Samples

Region	Urban			Rural	
	Regional center, 300,000+	*Large city, 100,000+*	*Mid-sized town, 50–100,000*	*Small town, 10–50,000*	*Rural settlements, fewer than 10,000*
Brest		Pinsk, 135,000 Baranovichi, 170,000	Kobrin, 52,000		
Vitebsk			Polotsk, 84,000	Tolochin, 10,200	Verhnedvinsk, 7,200
Gomel	Gomel, 505,000		Zhlobin, 80,000	Kalinkovichi, 38,000	
Grodno	Grodno, 350,000		Volkovisk, 44,000	Smorgon, 36,000	
Minsk	Minsk, 1,900,000	Borisov, 145,000		Vileika, 27,000	
Mogilev	Mogilev, 366,000	Bobruisk, 217,000			Krugloe, 7,300

Note: Four regional centers (300,000+), four large cities (100,000+), four middle-sized towns (50–100,000), four small towns (10–50,000), two rural settlements (fewer than 10,000).
Source: SATIO.

- Social benefits: Those in two bottom income quintiles were also divided into two subgroups: those who receive targeted social assistance (four FGDs) and those who do not receive these benefits (eight FGDs).
- Geography. The sample covered urban and rural areas in different regions:
- Gender: 12 FGDs included participants of both genders, three FGDs—with men and three FGDs—only with women.
- FGDs included participants of three different age groups: young people (21–30 years), middle-aged (31–45 years), elderly people (46–65 years).
- In addition, several subsamples of vulnerable population were included in the research (female pensioners living alone, families with many children (three and more), families with disabled children or parents, single mothers). Sub-samples' representatives of each subsample took part at different FGDs.
- Three discussions were held with household groups that have no access to district heating (separate houses of private sector which are heated individually, by the means of gas or wood).

In-Depth Interviews

Eleven in-depth interviews were conducted with representatives of district heating companies, targeted social assistance, and housing maintenance units administrators. The structure of the in-depth interview sample is summarized in table B.2. The FGD discussions consisted of five exercises dedicated to different aspects of the researched problem:

- Communal services and consumption patterns
- Coping with energy payments
- Social assistance benefits to support energy expenditures
- Awareness and attitudes toward district heating
- Awareness and attitudes to reforms in district heating sector.

Table C.2 Structure of In-Depth Interview Participants

1	Tree representatives of: Mogilev utilities management organization, district heating companies (Mogilevenergo, heating networks)	Mogilev
2	Representative of district heating company	Borisov
3	Representative of targeted social assistance	Borisov
4	Representative of district heating company "Bobruiskzhilkomhoz"	Bobruisk
5	Representative of housing maintenance unit	Volkovisk
6	Representative of targeted social assistance	Volkovisk
7	Representative of housing maintenance unit	Kohanovo village, Tolochin region
8	Representative of targeted social assistance	Tolochin
9	Representative of targeted social assistance	Pinsk
10	Representative of district heating company "Zhilkomhoz"	Krugloe
11	Representative of targeted social assistance	Krugloe

Source: SATIO.

Ethnographic case studies were performed with the representatives of vulnerable social groups, such as single mothers, elderly women, and families with many children. The interviews were conducted at homes of the residents, were devoted to the specific coping strategies they apply to deal with increasing public utility payments.

The Localities Chosen for Focus Group Discussions and In-Depth Interviews

The study was conducted in 18 settlements of Belarus in different areas of the country. Settlements differ by sizes, ranging from large regional centers to small towns and rural settlements.

Minsk

Minsk is the capital of Belarus, the administrative center of Minsk oblast and Minsk region, an independent administrative unit with a special (capital) status. Minsk is the major transportation hub, political, economic, cultural, and scientific center of the country. Minsk is divided into nine districts. Minsk population is about 20 percent of the total country population. "Minskenergo" (HPP-3, 4) and "Minsk heating networks" (Minsk HPP-2, 9 district boiler-houses, 208 central heat substations) are direct producers and suppliers of thermal energy to consumers in Minsk. "Minsk heating networks" and "Minskkommunteploset" provide maintenance services for heating networks in Minsk. And housing and communal maintenance services in Minsk housing are provided by "Minsk municipal housing and supply utilities."

Gomel

Gomel is an administrative center of Gomel oblast and Gomel region, the second most populous city in the country. The city is divided into four districts. Gomel is a city with developed industry, science and culture, an important transportation hub. The city has more than 100 enterprises of mechanical engineering, light industry, food, chemical, and other industries. Direct production and consumers supply of thermal energy in Gomel is provided by "Gomelenergo." Housing and communal maintenance services in Gomel housing are provided by "Gomel municipal housing and supply utilities."

Mogilev

Mogilev is a city in eastern Belarus, the administrative center of Mogilev oblast and Mogilev region. Mogilev is the third most populous city in the country. City is divided into two districts. Mogilev is one of the largest industrial and cultural centers of the country. The city has 70 industrial enterprises. "Mogilevenergo" (HPP-1, 2) and "Mogilev heating networks" (small boiler houses) are direct producers and suppliers of thermal energy to population in Mogilev. Housing and communal maintenance services are provided by both "Mogilev heating networks" (a half of the city) and "City water canal" (another half of the city).

Grodno

Grodno is a city in eastern Belarus, the administrative center of Grodno oblast and Grodno region. Grodno is a modern European city, a major administrative, industrial, scientific, and cultural center of the republic. Grodno is a huge industrial center of Belarus. Leading position in the industrial complex of the city and the oblast belongs to the largest petrochemical complex in Belarus "Grodno Azot." "Grodnoenergo" (HPP-2) and "Grodno heating networks" (North HPP, boiler house "Devyatovka") are direct producers and suppliers of thermal energy to consumers in Grodno. Housing and communal maintenance services in Grodno are provided by "Grodno municipal housing and supply utilities."

Pinsk

Pinsk is a town of regional subordination, the administrative center of Pinsk region in Brest oblast. The town has more than 50 industrial companies, the largest of which is the "Pinskdrev." Pinsk HHP and mini-HPP "Western" (which are on the balance of the "Pinsk heating networks") produce and supply thermal energy to consumers in Pinsk. Housing and communal maintenance services in Pinsk are provided by "Pinsk housing and supply utilities." In 2013, the state-targeted social assistance was provided to 1021 households of Pinsk region.

Baranovichi

Baranovichi is a town in Brest oblast, the administrative center of Baranovichi region. Baranovichi region is one of the largest agricultural producers in the Brest oblast. Structure of agricultural complex of the region is represented by more than 10 enterprises, the largest of which is the poultry factory "Druzhba."

Belarus Heat Tariff Reform and Social Impact Mitigation
http://dx.doi.org/10.1596/978-1-4648-0696-4

"Baranovichi municipal housing and supply utilities" and its subordinate organizations provide services for the supply of hot water and production and distribution of heat energy to the citizens and other consumers of the town.

Borisov

Borisov is a town in Minsk oblast, the administrative center of Borisov region. Borisov's second most important industrial city of Minsk region; there are 42 plants and factories. Unitary Enterprise "Zhilye" provides services for the supply of hot water, for production and distribution of heat energy to the citizens of the town (small part of the town), and for housing and communal maintenance services. And, the main producer of heat energy for Borisov is combined heat and power boiler of "Minskenergo" (situated in Zhodino).

Bobruisk

Bobruisk is a town in Mogilev oblast, the administrative center of Bobruisk region. The town is divided into two districts. There are 41 enterprises of different industries in the town. "Bobruiskzhilkomhoz" provides services for the supply of hot water and production and distribution of heat energy to the citizens of the town. The enterprise has 25 boiler houses and 81 central heat substations. Also, one of the producers of heat energy for Bobruisk is HPP-1 of "Mjgilevenergo" (provides heat to a small part of residential sector but mostly—to industry of the town).

Kobrin

Kobrin is a town in Brest oblast, the administrative center of Kobrin region. There are 22 industrial enterprises in Kobrin region, the vast majority of which are located in Kobrin. "Kobrin HSU" provides services for the supply of hot water and for production and distribution of heat energy to the citizens of the town. The enterprise has 30 boiler houses, 21 central heat substations. Nineteen boiler houses run on gas and 11 on local fuels. In 2013, the state-targeted social assistance was provided to 972 families in Kobrin region.

Polotsk

Polotsk is a town in Vitebsk oblast, the administrative center of Polotsk region, the most ancient town of the country. There are a number of industrial enterprises in Polotsk: wine factory, bakery, dairy plant. "Polotsk HSU" provides services for the supply of hot water and for production and distribution of heat energy to the citizens of the town. The enterprise has 11 boiler houses and 5 central heat substations.

Zhlobin

Zhlobin is a town in Gomel oblast, the administrative center of Zhlobin region, the third most populous town in the oblast after Gomel and Mozyr. The largest enterprise of the town and the country is "Belarusian Steel Works." HSU "Unicum" provides services for the supply of hot water and for production and distribution

Belarus Heat Tariff Reform and Social Impact Mitigation
http://dx.doi.org/10.1596/978-1-4648-0696-4

of heat energy to the citizens of the town. The enterprise has 44 boiler houses (24 boiler houses run on local fuels) and 26 central heat substations.

Volkovisk
Volkovisk is a town in Grodno oblast, the administrative center of Volkovisk region. There are 13 large industrial enterprises in the region. The main housing is multidwelling apartment blocks. "Volkovisk HSU" provides services for the supply of hot water and for production and distribution of heat energy to the citizens of the town. The enterprise has 26 boiler houses and 12 central heat substations. Twenty-one boiler houses run on gas and 5 on local fuels. And housing and communal maintenance services in town housing are provided by "Volkovisk housing maintenance service."

Tolochin
Tolochin is a town in Vitebsk oblast, the administrative center of Tolochin region. There are six large industrial enterprises in the region. The main housing is multidwelling apartment blocks and individual housing. "Tolochin-kommunalnik" provides services for the supply of hot water and for production and distribution of heat energy to the citizens of the town. The enterprise has 15 boiler houses in the town and in the region. There is a separate HSU with its own boiler houses in Kohanovo village.

Kalinkovichi
Kalinkovichi is a town in Gomel oblast, the administrative center of Kalinkovichi region. There are 10 industrial enterprises in the region including meat processing plant, dairy factory, and furniture factory. The main housing is multidwelling apartment blocks and individual housing. "Kalinkovichsky-Kommunalnik" provides services for the supply of hot water and for production and distribution of heat energy to the citizens of the town.

Smorgon
Smorgon is a town in Grodno oblast, the administrative center of Smorgon region. There are 10 huge industrial enterprises in Smorgon region. The main housing is multidwelling apartment blocks and individual housing. "Smorgon HSU" provides services for the supply of hot water and for production and distribution of heat energy to the citizens of the town. The enterprise has 20 boiler-houses and 10 central heat substations. Six boiler houses run on gas and 14 on local fuels. In 2013, the state-targeted social assistance was provided to 972 families in Smorgon region.

Vileika
Vileika is a town in Minsk oblast, the administrative center of Vileika region. There are 10 industrial enterprises in the town. The largest artificial lake in the country—Vileiskoye reservoir—is near the town. The main suppliers of heat to consumers are "Vileiskoye HSU" (there are town and region boiler houses on the

balance of the enterprise) and mini-HPP of "Minskenergo." Housing and communal maintenance services in the town are also provided by "Vileiskoye HSU."

Verkhnedvinsk

Verkhnedvinsk is a small town in Vitebsk oblast, the administrative center of Verhnedvinsk region. There are six industrial enterprises in the region: flax mill, dairy plant, bakery, and others. The main housing is multidwelling apartment blocks and individual housing. The main supplier of heat to consumers is "Vernedvinsk HSU." In 2013, the state-targeted social assistance was provided to 706 households in Verhnedvinsk region.

Krugloe

Krugloe is a rural settlement in Mogilev oblast, the administrative center of Krugloe region. Here are the department of Mogilev dairy plant, Krugloe flax plant, and other enterprises. The main housing is multidwelling apartment blocks

Map C.1   Geographical Distribution of Focus Group Participants

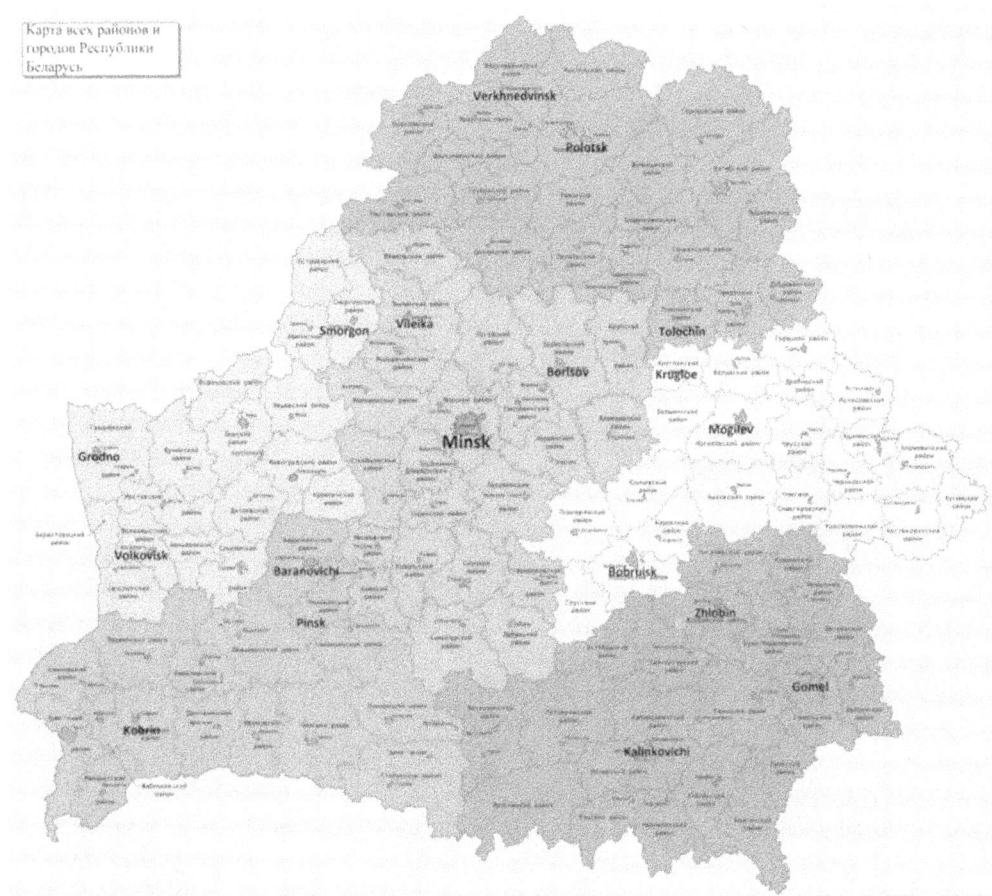

Source: SATIO.

Belarus Heat Tariff Reform and Social Impact Mitigation
http://dx.doi.org/10.1596/978-1-4648-0696-4

and individual housing. "Zhilkomhoz" in Krugloe provides services for the supply of hot water and for production and distribution of heat energy to the citizens of the town and the region. The enterprise has 17 boiler houses; 2 boiler houses run on gas and 15 on local fuels. In 2013, the state-targeted social assistance was provided to 359 households in Krugloe region.

Communicating Heating Tariff Reform to Household Lessons and Experience from Eastern European Countries and Russia

Poland

Background

Poland was one of the first countries among former Soviet bloc which raised utility tariffs for households. Poland (as well as Lithuania, Latvia, and Estonia) followed a 'Big Bang' approach (Terapia Szokowa) to reforms meaning that most prices were set to reflect cost recovery. Tariffs were raised in two stages: sharp rise in 1990–94 and the subsequent gradual increases in tariffs. Prior to tariff increases, the share of utilities in the total expenditures of households was 2.5 percent (working population) and 4.9 percent (the elderly), and at the same time Poland had a system of subsidized energy costs. After the introduction of the new pricing system for utilities, subsidies were eliminated within four years. Natural gas tariffs rose by more than 200 percent in 1991 and by more than 50 percent in 1992. District heating tariffs were raised sixfold by 1994.

Despite sharp increases in tariffs, the new tariff policy was not accompanied by concrete measures to protect the poor. Besides, social protection programs were not able to cope with increases in poverty due to improper planning and poor administration. To address this issue, the Polish government introduced a system of housing subsidies in 1995, but it did not succeed in resolving the problem. It is worth noting that according to the Polish laws defaulters were evicted from their apartments. Rigid rules of granting assistance and insufficient communication with households resulted in low coverage with only 6 percent of households receiving the subsidy. Consequently, the average household spent about 7 percent of total expenditures compared with less than 4 percent in Germany in 1997.

The second stage of utility tariff reform in Poland was regulated and administered more efficiently. Growth of tariffs slowed down and economic recovery in Poland allowed to benefit from positive dynamics of real incomes. The Polish government introduced more efficient system of housing subsidies and increased its communication with households.

This appendix was contributed by Irina Oleinik.

Communication Approach

Targeted public information campaigns were critical to ensure public buy-in with energy reform efforts and to shift behavior patterns toward energy saving. The key speakers were representatives of local governments and local communal service holdings (which includes energy supply companies). National authorities did not coordinate the information campaign but recommended to local governments to conduct these activities by themselves. Central government officials regularly appeared on TV and other media with updates on new legislation, social protection mechanisms, and changes in tariff policy. Communication campaign focused on explaining to households the nature of new tariffs and the components of the heating price. Following were the main messages: "Tariffs should compensate the price [may be cost, because the key point is cost recovery]of energy"; "Energy saving will reduce your invoice." To gain public support for heating tariff reform, municipalities conducted public hearings at the local level. Public hearings revealed the need for a clear and transparent mechanism for selecting enterprises to provide the services. Within the framework of the awareness raising campaign, officials met with civil society activists and communities' opinion leaders. No specific campaigns were organized in the media except regular contacts with mass media representatives (press conferences, interviews, participation in TV and radio programs).

Today, the Polish government conducts no communication activities related to heating tariffs (because the issue became almost nondisputable, for example, in 2007–10 growth of tariffs coincided with the consumer price inflation rate and remained almost unchanged). The key focus of the current government communication strategy is to promote energy efficiency and energy-saving practices among both industrial and residential consumers. It is worth mentioning the campaign on promoting energy efficiency behavior among schoolchildren (see planetaenergii.pl)

Hungary

Background

Hungary began to increase tariffs later than Poland, in mid-1990s. However, new tariffs were below cost-recovery levels. Subsidization was completely eliminated only at the last stage of tariff reform (2006–10). The program of housing subsidies was launched in 1993 and was intended to provide financial assistance to households for utility payments. Besides, in 1997–98 the temporary Social Energy Fund provided subsidies to the most disadvantaged families.

At the second stage of reform (2006–10), the government introduced an additional social protection tool—a special program for reimbursement of payments for gas. The program was administered by the State Treasury and covered the majority of households depending on their financial status. However, in 2011, the program was discontinued because it absorbed significant budget resources.

Communication approach: The Hungarian government made some aware-ness-raising efforts for publicizing social protection measures mostly during the second stage of tariff reform. In particular, the government officials conducted numerous consultations with local communities, disseminated information (leaflets on how to apply for reimbursement of payments for gas), and offered consultations both on an individual level and in the media (publications, replies to questions from the audience, participation in TV and radio shows).

Estonia

Background
Estonia followed the same approach to heating tariff reform (shock therapy) as Poland in 1992–96 when energy prices were raised every year. National legislation also provided for eviction of defaulters from their apartments, but the number of evicted families was very low compared to Poland.

Communication Approach
At the beginning of heating tariff reform, central government, particularly the Minister of Communal Property and the Minister of Finance, communicated the need for tariff adjustments, but later the focus of communication was shifted toward the local level with mayors and energy companies' representatives becoming the key speakers. The main messages centered on "Importance of energy metering, especially heat metering," "Importance of timely payment of heating bills," and "Energy saving." The most frequent communication channels included national and local media, information meetings with community members, individual consultations, and handouts with explanation of the need for tariff adjustments.

Bulgaria

Background
Bulgaria pursued more gradual heating tariff reform. In 1997–2005, prices rose by 10 percent annually with eventual elimination of subsidization. In winter 1996–97, the Government (with support from the European Union) launched the Program of Winter Surcharges (PWS). The program provided assistance for payment of utility services during the winter season on the principle of Incorporation of income for low-income households. Initially, the program was cofinanced from local budgets. However, given that local budgets did not provide sufficient funding, the program faced 30 percent deficit in financing and failed to ensure sufficient protection to households. The program funding entirely from the state budget has helped solve that problem since 2003 onwards. In parallel with tariff increases, the Bulgarian government implemented energy efficiency measures. The World Bank noted the high efficiency of targeted assistance of PWS after the problem of financing was resolved. In 2007, 70 percent of the PWS funds were received by 10 percent of the poorest households.

Belarus Heat Tariff Reform and Social Impact Mitigation
http://dx.doi.org/10.1596/978-1-4648-0696-4

Communication Approach

The Bulgarian government focused its communication effort on encouraging households to save energy. The government made a review of the structure of the average household expenditures in order to identify acceptable level of tariff increases. Broad communication campaign was supplemented by parallel administrative measures on introduction of personal responsibility for monitoring of energy consumption. The Ministry of Energy and Energy Resources launched the campaign for installation of new electricity meters in every household. All local service providers were obliged by the government to establish consulting centers responsible for dissemination of information materials, conducting information events and individual consultations. Nongovernmental organizations (NGOs; including consumers associations) were also actively involved in communication activities. Key speakers of the campaign included senior government officials (Minister of Energy and Energy Resources, his deputies) and mayors. The main messages included "Responsibility of household to pay for energy on a regular basis," "Energy saving practices," and "Subsidies are transparent and targeted only at families in need." The key issues covered in outreaching efforts included breakdown of tariffs and energy-efficient practices. As a result, the Bulgarian government succeeded in introducing an efficient targeted social protection program and in encouraging adoption of energy-efficient practices by households.

Russia

Communication Approach

Russia has extensive experience in explaining consumers the composition and calculation of tariffs (the Federal Service of Tariffs has a well-designed website http://ftstrf.ru) and in communicating energy efficiency issues to children. For example, the NGO "Foundation for Facilitating Utilities Sector Development" has developed and introduced an online game for kids "Jeka" (igra-jeka.ru), which explains how to save energy at a household level.

Lessons Learned

Communication of heating tariff reform should be pursued in parallel with dissemination of information about measures taken by the government to safeguard protection of the poorest households from adverse effects. It is particularly important to demonstrate to the public that these measures are properly targeted and efficient.

- Broad outreach program needs to be designed to address the information gaps and to actively provide information to the population at large to inform them of energy-saving opportunities. Once behavioral changes have taken place and a "tipping point" established, the resources needed to maintain this impact are reduced.

- Partnerships are vital—experience from other countries revealed that it is essential to involve local authorities, NGOs, consumer associations, and community leaders in communication efforts.
- Countries are increasingly making use of communication channels beyond traditional media, such as Internet and educational institutions (schools and universities).

Belarus Heat Tariff Reform and Social Impact Mitigation
http://dx.doi.org/10.1596/978-1-4648-0696-4

References

Energy Charter Secretariat. 2013. "In-Depth Review of the Energy Efficiency Policy of the Republic of Belarus."

Grainger, Corbett, Fan Zhang, and Andrew Schreiber. 2015. "Distributional Impacts of Energy Cross-Subsidization in Transition Economies: Evidence from Belarus." World Bank Policy Research Working Paper, World Bank, Washington, DC.

ECO-AUDIT

Environmental Benefits Statement

The World Bank Group is committed to reducing its environmental footprint. In support of this commitment, the Publishing and Knowledge Division leverages electronic publishing options and print-on-demand technology, which is located in regional hubs worldwide. Together, these initiatives enable print runs to be lowered and shipping distances decreased, resulting in reduced paper consumption, chemical use, greenhouse gas emissions, and waste.

The Publishing and Knowledge Division follows the recommended standards for paper use set by the Green Press Initiative. The majority of our books are printed on Forest Stewardship Council (FSC)–certified paper, with nearly all containing 50–100 percent recycled content. The recycled fiber in our book paper is either unbleached or bleached using totally chlorine free (TCF), processed chlorine free (PCF), or enhanced elemental chlorine free (EECF) processes.

More information about the Bank's environmental philosophy can be found at http://crinfo.worldbank.org/wbcrinfo/node/4.

green press
INITIATIVE